How *Smart* Is Your **Dog**?

How *Smart* Is Your Dog?

Discover if your pet can solve
these fun canine tests

David Alderton

Quercus

Contents

Introduction

Acquiring a puppy—or an older dog—is an exciting event as you welcome the newcomer into your family. However, then the question arises: how will your pet get on with the training needed to master basic skills? This book is a unique guide to help you chart your pet's progress and to see just how smart it is, and it also gives you plenty of tips to help you achieve the best training results. But do bear in mind that some types of dogs will find it easier to perform certain tasks than others.

AIMING FOR THE STARS

The good news is that this is not a straight pass or fail situation. Instead, if at first you don't succeed, keep trying and see if things have improved after a further month or two. The unique scoring system lets you monitor your pet's progress easily over time. Record the stars using

★ = getting to grips with the basics of simple training

★ ★ = real improvement as your dog masters new tasks

★ ★ ★ = ability and proficiency; your pet is a star pupil

Puppies love to explore their surroundings and are keen to learn, which means you need to be a good teacher.

can teach your dog, and you may even decide that you then want to take part in various competitions. An ever-increasing number of canine sports are now being recognized. Alternatively, you may want to teach your pet a range of tricks and then film these to post on YouTube and similar Internet sites.

Left: Bear in mind that some dogs, like the Shar Pei for example, are harder to train than other breeds, so be prepared to be patient if necessary.

Below: It's a cute trick but it can also benefit your dog in the future. Persuading your pet to let you lift a paw helps with treatment like bandaging, if it cuts a pad.

the charts at the end of each section and see your pet's overall score on page 187.

It is vital that your pet masters a number of basic lessons, so that it will be well-adjusted to household living and able to have fun safely. These basic lessons include coming when called, not running away, and being well-disposed toward other dogs, as well as to family and visitors. There are also lots of fun games and activities that you

PRACTICAL POOCH POWER

Dogs are great! They help us in so many ways, and some remarkable findings are coming to light regarding the benefits of dog ownership from a scientific perspective. Dog owners tend to be healthier overall than the average population, with better recovery rates from heart attacks. Older people also worry less about being burgled or becoming the victim of crime if they have a dog living with them.

Above: Dogs are playing a key part in reading schemes, with children being more willing to read aloud to a dog than to their own peers.

Left: Dogs come in all shapes and sizes, so there is bound to be one that appeals to you!

If you want to meet a new partner, it may well be that getting a dog turns out to be a better option than joining a dating agency! This is because strangers are up to four times more likely to start talking to someone walking a dog in a park than they are to strike up a conversation with a total stranger in the same place. So your dog may be just the matchmaker you're looking for!

Dogs also play an important role in helping people who have disabilities or are ill. They are now used as assistance dogs in a growing number of areas, being best-known for aiding the visually-impaired as well as the deaf. Dogs now work with epileptics as they are able to warn of the likely onset of seizures, and they also help diabetic patients as they can sense a catastrophic falloff in blood sugar levels that can lead to unconsciousness.

Whatever dog you have, it can be both smart and excellent company, so enjoy a relationship that will bring out the best in both of you!

Even if you've had a bad day, being greeted by your pet dog will always cheer you up.

Amazing senses

The scenting ability of dogs, which is estimated to be about 10,000 times more sensitive than our own, means that they are now being trained to help detect various types of cancer without the need for invasive procedures. This entails them testing urine samples by smell for early indicators of prostate cancer. The initial results of these tests are very encouraging.

How Your
Dog Learns

Superdog or Doggy Dunce?

Don't rush into making any snap judgments about your pet's brain power. You could end up getting a surprise, because when it comes to training your dog correctly, a lot actually depends on you. Dogs are generally eager to learn, but you need to teach them in a consistent way that they can easily understand.

1 Bear in mind that puppies usually learn faster than adults. They are programmed to pick up information quickly at an early age, which helps their chances of survival in the wild.

2 Keep sessions short to maximize your dog's concentration span. If training stretches on for more than 5–10 minutes, your dog is likely to lose interest and become distracted.

3 Make learning fun! Be enthusiastic and always encourage your pet, rather than scolding him if he fails to pick up what you want. Praise and reward are great motivators.

4 It helps to build up set training routines in a step-by-step fashion, so that you can go back and repeat sections that need a refresher while still teaching your dog new things.

★ Your dog is an enthusiastic pupil

★ ★ He understands the basics but needs more practice

★ ★ ★ He learns new things quickly

When dogs are in unfamiliar surroundings or particularly excited, they sometimes seem to forget things that they have already learned, so you may need to make allowances on these occasions.

Blame the teacher!

If your dog is proving slow to learn, you may be the problem! Perhaps you are expecting your pet to learn too quickly, and he is becoming confused by your instructions? You should aim to be as consistent as possible during all your pet's training sessions.

Genes and Genius

Dogs have been bred to undertake an amazing variety of tasks over the centuries, and these talents are reflected to a greater or lesser extent in their breed characteristics. If you are looking for a canine athlete, then clearly the speedy Greyhound would outshine an ambling Bulldog. The scenting skills of Bloodhounds are unsurpassed, but if you want a dog with keen eyesight, then an Afghan Hound is an ideal choice. If you are looking for a lovable companion to sit on your lap, a Chihuahua comfortably beats a St. Bernard!

1 Dogs with working ancestries tend to be easier to train than those reared exclusively for show purposes. This may reflect the environment where they grow up.

2 Nature versus nurture. Puppies undoubtedly learn by watching older dogs, and so correct training is probably more important overall than your puppy's breed ancestry.

The Lagotto Romagnolo is a highly specialized Italian dog breed that can sniff out valuable truffles growing hidden underground.

Trained to fetch

Although traditionally used for retrieving game, Labrador Retrievers can easily be trained to bring you specific objects from around the home.

★ Your puppy recognizes you

★ ★ He responds as his breed characteristics suggest

★ ★ ★ He reacts in accordance with his training

3 If you do decide to go to training classes with your pet, remember that your dog will probably learn better from you than from a stranger.

Below: If a puppy appears unresponsive to your calls, he could be deaf. This is expecially true of Dalmatians.

Fit for Fun

The way your dog is physically built will have a bearing on what he can be expected to achieve. Breeds have evolved and been selectively adapted for a host of different purposes.

Breeds with short muzzles, such as the French Bulldog, will not be natural athletes.

1 Some types of dogs are sprinters, bred to race at speed, whereas others are more like endurance runners, able to display considerable stamina over long distances.

2 The way a dog functions depends on its "conformation"—how it is put together. A deep chest, giving good lung capacity, and strong hind legs add up to create a powerful runner.

3 Short-legged breeds do not have the stride length of bigger dogs, which means they cover less ground at each step. Consequently, for them running requires lots more energy.

Getting into their stride
Some hounds exist in standard and basset ("low" legged) forms. Longer-legged varieties accompany riders on horseback; basset packs are followed by huntsmen on foot.

True or False?

As well as Greyhounds, another breed that has been raced competitively around a track is the Afghan Hound.
(Answer on score chart, p61)

☐ True

☐ False

Basset Hounds may not be fast, but they can follow scents determinedly over long distances.

Bred to run, Whippets used to be nicknamed "the poor man's racehorse," and they were raced in streets in the northeast of England.

Man's Best Friend

It is no coincidence that certain types of dog are often selected to work closely alongside people. This is generally because they prove more responsive to their handler, and their willingness makes them easier to train. Breeds in this group include Labrador Retrievers, Border Collies, and the various types of shepherd dogs.

A Labrador Retriever puppy. This is the most popular breed in the world today, thanks largely to its easygoing and adaptable nature, combined with a natural readiness to learn.

POPULAR CHOICES

1 All these breeds were developed originally to work on a one-to-one basis with people—not as pack dogs.

2 They respond well to training, but they also possess sufficient intelligence to adapt and act on their own initiative.

3 These breeds of dog are typically observant, which not only helps them to learn, but also to keep them focused when working.

4 Crossbreeds can also score well— notably Labradoodles, which are created by crossing Labrador Retrievers and Standard Poodles.

★ Your puppy settles well into his routine

★ ★ He masters basic training quickly and easily

★ ★ ★ He enjoys special training exercises

A German Shepherd Dog and a Border Collie—both breeds that have an ancestry of working with farm stock. Now their working abilities are being more widely utilized for our benefit.

Boredom busting

Dogs that are used to working need to be kept active. If this doesn't happen, they can become bored and destructive around the home. They also require space outdoors so they can get plenty of daily exercise.

Brain Games

A well-trained dog not only seems more intelligent than one that has not been taught the basics, but he will also be more confident and outgoing, because he has worked closely with you to reach this standard. Achieving success will encourage him to be even more resourceful and self-confident.

1 Try teaching your dog the name of each of his toys, and then see how quickly he learns to recognize them.

2 Next, leave some toys around the home, and encourage your pet to find the particular one you have asked for.

3 Carry on in this way, and see how many different items your pet can identify and recall, and then bring them to you on command.

On the shelf

You can now buy a whole range of toys that are intended to interest your dog and stimulate his intelligence. Some are designed to be more puppy-friendly.

★ Your dog recognizes the name of his toy

★ ★ He will bring this toy to you when asked

★ ★ ★ He can recognize more than one toy by name

Above: "Fetch the blue ball."
You'll get a real kick when your
dog brings you a named toy that
he has recognized simply from
your words.

Right: Part of the secret when
training a puppy or an older dog
is to maintain his focus on you.
Hand gestures can be useful to
keep him alert.

School of Fun

Dogs genuinely appear to enjoy learning, and they can become very good at it once you have established what is required. Leadership is very important to get the best out of your pet.

1 Once your dog has learned what is expected, you then want to encourage him to start focusing more closely on the activity. This in turn should assist his problem-solving skills.

2 Having fun—at least in your dog's mind—is not just about pleasing you with his reactions to a particular task, but may involve him following his instincts and acting independently.

3 Such behavior can cause problems, rather than being a source of satisfaction. Your dog may use his ingenuity to escape from your yard, for example.

4 Create activities for your dog that are fun for both of you and that help to reinforce aspects of his training.

Puppies are often mischievous and playful so try to nip any unwanted habits in the bud, as they might cause you problems as he grows up.

Dogs love their favorite toys, so aim to use them as training aids. Lessons that are fun are likely to be the most productive.

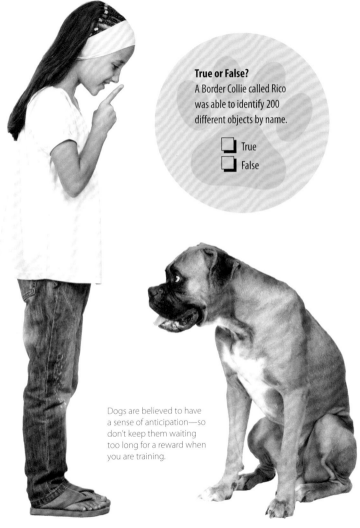

Jealousy may pay!

According to research, dogs may experience emotions such as jealousy. If one dog is being praised, for example, this may encourage his companion to perform better in the future.

Dogs are believed to have a sense of anticipation—so don't keep them waiting too long for a reward when you are training.

How Dogs Learn

Three components are thought to make up the canine learning process. Firstly there is instinctive intelligence, which allows the dog to carry out the tasks for which he was bred, then adaptive intelligence relating to his problem-solving skills, and finally the dog's ability to learn from us.

Do not be tempted to rush the basics of training—it'll be counterproductive. Take your time and build on each positive step forward.

1 The top dogs in the canine intelligence rankings are those that learn quickly, and are able to grasp what is required of them more or less from the first moment of training.

2 It is no surprise that when the test results returned by judges were compiled, the top dog breeds displayed high levels of adaptive intelligence, indicating that they are good at solving problems.

Smart chart

There have been various attempts to find ways to assess canine intelligence. The first chart of its type, compiled by psychology professor Stanley Coren and published in 1994, was based on judges' ratings of some 79 breeds, with the results standing the test of time. The Border Collie topped the list, to be followed by the Standard Poodle, then the German Shepherd Dog, Golden Retriever, and Dobermann Pinscher. In terms of the smaller companion breeds, the Papillon ranked highest at number eight, with the working breeds clearly outperforming this group.

In intelligence tests Golden Retrievers followed new instructions after no more than five repeated requests, but it may take up to 40 goes for a Bulldog to pick up the same message.

3 Ease of training is another important factor in learning. Dog breeds that were assessed as being the hardest to train were also ranked as being less intelligent, reflecting the fact they were less adaptable.

★ Your dog recognizes what he is meant to do when asked

★ ★ Your pet uses his initiative to solve a problem

★ ★ ★ He can be relied on to perform a task well

Smart Puppy

Is it possible to figure out whether one puppy in a litter is smarter than his littermates? This is not straightforward, but nothing ventured, nothing gained! One of the best signs to look for is an adventurous puppy who is keen to explore his surroundings.

Don't disturb the litter when you first see them, but watch them as they move about, and see which one looks most independent. This can be a good guide that he's going to grow up smart.

Below: Out for the count. Puppies get tired very quickly, so don't discount a sleepy puppy. He might simply be the most inquisitive member of the litter.

WHAT TO BEAR IN MIND

1 Some breeds are better-suited to learning from people than others, having been developed to work closely on a one-to-one basis with their owner. Do your homework!

2 See how the puppies react to you. Is one more curious and prepared to approach you? This is an encouraging sign, suggesting a bolder nature.

3 A recently-weaned puppy should learn much faster than an older dog, appearing smarter as a result, as well as bonding more rapidly with you.

Being smart in the wild is all about survival. A puppy learns from you in the same way that he would from other members of his pack.

Learning fast
Curiosity is an important characteristic when you are seeking a smart canine companion. Bear in mind, however, that breeds have been developed for different purposes, and some will instinctively learn what is required faster than others.

★ Does your puppy know his name?

★ ★ Does he know when it's dinner time?

★ ★ ★ Does he know when it's time for bed?

Quick to Learn

Puppies learn fast, reflecting the fact that in the wild they would have to deal with a hostile world where they need to grow up quickly. Survival skills are paramount in such circumstances. A puppy will look toward you for guidance at this critical time.

1 Don't always assume that your dog is fully motivated, or loves the treats on offer. If boredom strikes, be prepared to change your pet's training routine.

2 Dogs pick up on vocal commands, gaining an understanding of what is required by way of a response. It is the distinctive sound that is important, rather than any specific word.

3 They can also recognize and respond to hand signals, whether these are being used to reinforce verbal commands or employed separately from a distance.

4 Hand signals do need to be taught at relatively close quarters to begin with. You can increase the gap between you and your pet as he becomes more proficient, by backing away.

Young dogs love to explore the world around them and, as well as sniffing at an object, they may start to gnaw it. This behavior also helps to ease the pain of emerging teeth in their tender mouths.

Learning can be very tiring! Sometimes a puppy may fall asleep in the middle of a lesson.

★ Your puppy shows interest in learning

★ ★ Your pet's training is advancing in some areas

★ ★ ★ His training segments are linking in a sequence

Teenage kicks!

There is a great deal for young puppies to learn about and, at first, they tend not to stray too far from familiar surroundings. However, from about five months onward, hormones kick in and they begin to become more unruly and liable to "adolescent" misbehavior.

Puppies will make up their own games but, unfortunately, they sometimes involve chewing up a sneaker or a shoe.

Old Dog—New Tricks?

Contrary to the popular saying, it is possible to teach an old dog new tricks! But an aging dog will lose a little mobility, which makes some activities difficult, so be realistic in your goals. With an older newcomer in the house, first find out what he already can do reliably and well.

1 With an older dog, you must first win your new pet's confidence. Don't let him off the leash until you are sure he will come back when called.

2 Try to discover as much as possible about how your new dog has been trained previously, and the key words and signals that he has learned to respond to.

3 Swimming provides great exercise, particularly if your dog has painful joints, and it keeps his muscles toned. Your pet can learn to chase balls in a hydrotherapy pool.

Less active tricks, such as shaking a paw, can be taught quite easily in a dog's maturer years. They are not physically demanding.

It's vital that older dogs stay active. In a fetch-the-ball game, train your dog to drop the ball into your hand again, so that you can continue to play the game together.

4 Mental stimulation can be increased by encouraging scenting skills through games that allow your dog to use his nose to track down tidbits, even if his eyesight is failing.

★ Your dog shows mastery of lessons learned earlier in life

★ ★ Your pet can adapt some of his existing skills

★ ★ ★ He displays new skills learned in old age

Keep going!

Even if you have had an old dog from puppyhood, try to teach him new things to keep his mind alert and his body active. You may have to be more patient with your elderly pet, but your sense of achievement when he masters a new skill will be really satisfying.

Mischief Makers

Sometimes you seem to be making good progress when training your dog, but then you experience some apparent setbacks. This is not always your fault though, and successful training is not a constant upward curve. Some days will prove to be better than others—just keep working at it!

TROUBLE AHEAD

1 As dogs grow older, rather like teenagers they become more headstrong and determined. Their natural instincts to roam become stronger too.

2 Boredom can often be the cause of why your pet is not performing as he should, so vary his training routine.

3 Increased destructiveness around six months old may be linked to teething. Chews can help to soothe the pain of sore gums.

4 Male dogs especially may grow more disobedient as they become sexually mature, between six and 12 months of age, so neutering may then be advisable.

Be patient and make sure you communicate your wishes clearly. Remain consistent in your approach and accept that your growing dog wants to explore more. In time, the bond between you will become stronger.

★ Your dog is sometimes destructive or runs away

★ ★ Your pet begins to be more responsive to training

★ ★ ★ He doesn't cause damage and won't run away

Above: Attention-seeking puppies can be persistent. Dogs are very adept at learning how to get a response from their owners.

Right: A dog must be given more exercise as he matures. This will counter the effects of boredom which can lead to bad behavior.

Doggy distractions

The most common reason why things occasionally don't go well during training is that your dog is distracted. This can be for a variety of reasons, not all of which may be immediately obvious, such as the sound of thunder in the distance or a neighbor's dog barking.

Boredom Busters

Although a puppy will misbehave on occasions, it is important to prevent bad behavior becoming habitual.

A puppy will often become bored if he is left for a long time on his own, so try to avoid this situation. Think, for example, about taking your pet to a puppy creche if there is one near you.

Puppies play together regularly and keep one another amused when they are growing up as members of the same litter.

1 Accustom your puppy to being left alone for short periods of time. Play a game or take your pet for a walk beforehand, so he will be inclined to sleep in your absence, rather than looking for something else to do.

2 If your puppy starts to chew up shoes, make sure that none are left within reach around the home and provide some chew toys instead. If these are easily accessible, they will keep your dog happily occupied.

Above: A bored dog is likely to invent his own games when left alone and, inevitably, these often have a destructive element.

Below: Toys can help to substitute for littermates up to a point, but puppies require regular play sessions to keep their minds and bodies occupied.

Timing is critical
There is no point in scolding your pet, unless you catch him in the act of misbehaving, because he will not understand what he has done wrong after the event.

★ Your dog sometimes chews your things when left alone

★ ★ He will usually play with toys when he is on his own

★ ★ ★ Your pet's destructive behavior ceases

Dog's Eye View

Dogs see the world differently from ourselves, literally as well as in a more general sense. This is because their senses are not tuned in the same way as ours, which rely heavily on sight.

1 Dogs have much better nighttime vision than we do. They are able to see things clearly under conditions that appear as total darkness to humans.

2 Their view of the daytime world is not as bright as ours, because dogs are not able to distinguish colors as effectively as we can.

Left: A dog's vision naturally fades with age, but even a blind dog can find his way around thanks to the acuteness of his other senses.

Above: When looking into the distance, a dog's eyes are attuned to pick up movement, rather than focusing on stationary objects.

3 Dogs tend to have a wider view of the world than us, because of the position of their eyes in their heads. This is especially true with sighthounds.

4 At close quarters, dogs use their whiskers, which are specialized hairs, to determine if they can get their head and body through a narrow gap.

Acute hearing
Dogs have much better hearing than we do, and you can use this to help communicate with your pet. They can hear ultrasound frequencies, and if you buy a whistle that emits sound in this range, your pet will be able to hear it from a long way away and respond to your recall commands.

True or False?
Dogs are red-green color-blind, but see blueish and yellow colors clearly.

☐ True
☐ False

Dogs often enjoy watching the world passing by, but do not be surprised if your pet starts barking unexpectedly. He may have heard other dogs making a noise in the distance that you are unable to hear.

Teaching the Basics

The Name Game

You need to teach your dog to recognize her own name, and possibly those of other family members. But to avoid confusion, don't give your dog the same name as anyone else in the family. Always use your dog's name when calling or speaking to your pet, and she will soon learn the sound.

1 If you are stuck for a dog's name, there are lists on the Internet and even books to which you can refer. It must have a clear and a distinctive ring to it.

2 Choose a name that you like—you will be using it a lot. A short, uncomplicated one- or two-syllable name can be easily picked up at a distance by the dog. Don't choose a name that sounds like one of your obedience commands.

3 If you acquire an older dog that is used to her name, don't confuse her by changing this name when she comes to live with you.

It is important that your dog learns to respond to her name.

Name calling

It takes time for a young dog to learn the sound of her name. This will require continued repetition. Every time you want to get your dog's attention, start the sentence by using her chosen name.

★ Your dog reacts to her name

★ ★ She responds to instructions given by name

★ ★ ★ She will find you in another room when called

It's unlikely that your dog will recognize her name as a "name," but she will certainly identify with the sound of it.

Dog Days and Nights

Dogs are very much creatures of routine, and are happiest if they have a set structure to their days. They know what to anticipate at certain times, and use a variety of cues to detect the passage of time through the day.

1 It helps the training process too if you develop a routine with your pet. Keep the lines of communication simple, and she will soon learn what is expected.

2 Training your pet to be clean in the home will initially involve you in taking her outside first thing in the morning, after meals, and at night.

Conditioning
Although we might think that we control our dogs, they readily tune in to clues in our body language, like picking up a leash before a walk, and then turn these to their own advantage when they are guaranteed our attention.

3 It will take at least six months before your dog starts to ask to go outside to relieve herself, so in the meantime stay alert to the risk of accidents!

Dogs soon come up with ways to remind you if you are late with their food.

Dogs come to recognize the time of day when they are taken out for walks, although they cannot tell the time as such. Clues such as you reaching for a coat or a pair of boots can serve as the triggers.

★ Your dog is keen to be fed at mealtimes

★ ★ Your pet tells you when she is ready for her food

★ ★ ★ She lets you know when she wants to go outside

Puppies benefit from an established routine. Always take a puppy outside after she has eaten, as this is when she is likely to want to relieve herself. It will help to prevent accidents in the home.

Chilled Out

Is your dog generally happy and relaxed? Your pet can communicate her mood very clearly, if you learn to interpret the signs. It is certainly helpful if you can understand what your pet's body language is saying to you.

1 You will soon recognize when your dog is relaxed. This will usually be after a walk or a meal, when your pet will be ready for a snooze.

2 It helps your dog to learn to settle down if you take her for a walk before you go out, as she should be happy to go to sleep when she returns home.

3 As dogs grow older, they are likely to sleep for longer. However, they are usually always ready to go out for a walk if given half a chance!

Your dog will often sit down beside you when ready to rest, and may even paw at you gently to be made a fuss of, encouraging you to stroke her fur.

Dog doze

Dogs may spend as much as 14 hours asleep every day, although they wake up during this time if something interesting is happening. Large breeds like St. Bernards tend to sleep for longer than smaller ones.

★ Your dog sits down after exercise

★ ★ She lies down when she is tired or relaxed

★ ★ ★ She is totally chilled and rolls on her back

Chilled out and resting, this young puppy is lying on her back in the hope of getting her tummy stroked.

Young puppies tend to have periods when they run around burning up energy, and then they stretch out on their beds and fall asleep.

Walkies!

Your dog will be super alert to the tell-tale indicators, even before you have said anything! A rustle of a coat, the sound of shoes being changed, or the distinctive noise of a leash and collar will all serve to alert your dog that a walk is on the cards.

It is easier to train large, strong dogs when they are small puppies. You do not want an adult dog of this size running off ahead of you or pulling strongly when she is on the leash.

Extendible leashes are useful in places where dogs cannot be let off for a free run.

1 When the leash is first attached, a puppy may roll over on her back, trying to break free. This behavior is quite normal, and will soon pass.

2 Your puppy must learn to walk in a straight line, and should not pull ahead. Walking alongside a fence is good training for this discipline.

3 Introduce "stop and sit" into the walking routine. These different actions add variety and interest for your dog as she learns the ropes.

4 Monitor your pet's training progress. Dogs learn at different rates. Keep sessions short to ensure your dog maintains good concentration.

Walkabout

Although regular walks will obviously keep your pet fit, they also help to meet a dog's need for mental stimulation, which prevents boredom. There will be lots of interesting scents to investigate, and meetings with other dogs to enjoy.

★ Your dog accepts a collar and leash without struggling

★ ★ She walks on the leash but is inclined to pull

★ ★ ★ Your dog walks to heel and doesn't pull ahead

Although puppies cannot be taken out safely until they have completed their inoculations, they can be familiarized with their collar and leash by practicing around the house.

His Master's Voice

Your tone of voice is very important when it comes to communicating with your pet. The impact of using a harsh voice, showing that you are not pleased, will soon be apparent. Your dog will rapidly come to recognize whether you are satisfied—or not—as her body language will confirm.

ATTENTION! ATTENTION!

1 Whenever you communicate with your pet, you should start by getting her attention. Either call her by name or, if she is some distance away from you when you want her, use a whistle.

2 Dogs react best to a positive, encouraging tone of voice. If she fails to follow instructions, don't scold her—this is counterproductive. Instead, repeat the instruction clearly; you are trying to encourage her to get things right at the next attempt.

3 Scolding in a harsh tone of voice should be reserved for the occasions when you actually catch your dog misbehaving, perhaps chewing up something that she should not have, digging in a flower bed, or stealing food from the kitchen.

This dog knows that she is in trouble, as shown by her hangdog expression, with big eyes and the ears lying flat.

Left: Dogs may jump up when they become excited. Don't overreact but simply say "down" firmly, as you lower the dog's front legs back to the ground.

Below: Certain breeds, such as pointers, can be especially sensitive to their trainer's voice. This is probably because they develop a particularly close relationship with their handler.

True or False?
His Master's Voice was a record label named for a painting of a dog called Nipper listening to a gramophone.

☐ True
☐ False

Follow my leader
A young puppy will be looking to you for guidance. Although she does not understand the words, she will recognize whether you are pleased or not from the tone of your voice.

Sitting Comfortably?

Sitting upright is more comfortable for certain types of dog than others. Long, narrow-bodied hounds such as the Borzoi often prefer to lie down—either on their chests or their sides—rather than sitting on their slender haunches.

1 Successful training is based on linking a series of different simple instructions together, so that you can build up fluent and complex actions from smaller individual elements.

2 Once you have your dog's attention, apply gentle pressure to her hind quarters as you say "sit." Once she has adopted this posture, remember to praise her and reward with a treat.

3 Start by encouraging your dog to sit each time before putting her food bowl down. She will soon learn this routine as food is always a powerful incentive to learn good behavior.

Sitting is a simple lesson for your dog to learn, and it needs to be one of the first basic instructions that your pet masters. It's important because it allows you to gain your dog's attention.

★ Your dog seldom sits when requested

★ ★ Some of the time your dog sits when you ask her

★ ★ ★ Your dog consistently sits when she is asked

Never take the food bowl away if your dog does not sit instinctively at first. Simply repeat the command and encourage the sit by moving the bowl back over her head.

Sitting pretty
Most dogs will instinctively sit, even as young puppies, so you need to encourage them to do this when instructed on your verbal cue. Link the word "sit" to this action from an early age.

Training your dog to sit results in a much more relaxed domestic environment. Even the youngest members of the household can become involved in the fun of training this skill.

Sit and Stay!

Learning to sit and stay until given the cue to move is a vital part of a dog's training. Mastering these commands could even save your pet's life if you find yourselves in a situation where your dog could otherwise be in danger, such as running off the leash toward a busy road.

1 To build this combination, your dog must first have learned to sit without hesitation. You then need to add the "stay" component as the next step.

2 Always choose somewhere that is quiet and free from distractions for training purposes. Put your dog in a sit. Tell her to "stay" and back away a few paces. If she remains in position, praise her and reward with a treat.

3 Keep training sessions relatively short, and be patient. It is much better to repeat these several times a day, rather than going for one marathon training session at the weekend.

At first, your dog will probably stand up and try to follow you. She must be persuaded to stay.

Left: With your dog sitting alongside you, take a few steps away, while hopefully your dog remains obediently in a sitting position.

Above: Hand signals are very useful in this situation. A raised hand indicates that your dog should stay until instructed to come.

Retriever training

Gundogs obviously need to be very well trained before they start working with the guns. Having mastered sit and stay, they must then be taught a third command—fetch—to become proficient as retrievers.

★ Your dog sits readily when instructed

★ ★ Your pet sits but is reluctant to stay

★ ★ ★ She sits and stays patiently until asked to move

Total Recall

Coming when called is another potential life-saving lesson that your dog needs to be taught. Danger can arise in the most unexpected situations. You need to have effective control of your dog at all times, and particularly when you are walking together near roads or in unfamiliar countryside.

1 Start training your puppy to come back to you at an early stage, simply by calling her. Do this before meals, so that your pet starts to realize that obedience brings a great reward.

2 When you are playing ball, build on the opportunity to get your puppy to return to you when called to "come." This is quite easy, as the puppy soon learns that she needs to bring the ball back to you for the game to go on.

Harness your puppy's natural enthusiasm to come back to you when called, as she will not want to miss out on the possibility of a treat or praise and a welcoming stroke.

Above: Dogs are not always aware of danger, and their eagerness to see what's happening can take over.

Above right: Your dog could be running into danger, with a fast-flowing river or busy road in her path.

3 When you are out walking in woodland, you can turn coming back into a game of hide-and-seek that reinforces the recall message. Stand behind a tree trunk and call your dog back to you. She must seek you out to find you among all the other trees.

★ Your dog sometimes comes to you when called

★ ★ Your pet usually returns on command

★ ★ ★ She always comes promptly and obediently

Fun Times

Life's a lot more fun with a well-trained dog! And this applies just as much to the games you play together as to the more formal obedience training that underpins good canine behavior. Happy playtime builds a great bond between you.

HAVING A BALL

1 Remember that each dog is an individual, and they may like one type of game more than others. They are also likely to have favorite toys that they enjoy playing with.

2 Most dogs like to chase after balls, which they can catch and bring back to you. Other types of throw toys which bounce in unexpected directions are also good for games of fetch.

Let the good times roll! You will both feel energized and happy after playing games together.

3 Do not overtire your dog, or encourage her to play boisterously when the weather is very hot. Dogs cannot sweat as we do, and your pet may end up panting in a distressed state.

4 Keep an eye out for other dogs in a public space. They may want to muscle in on the game which can lead to conflict with your dog over the toy.

Don't be a spoilsport! You must let your dog win the game sometimes, or she will become frustrated. The mark of good play is when you are both having fun and are keen to go on.

Playtime!
Play seems to be an instinctive reaction for dogs. Even rescue dogs that have not been used to playing with toys will quickly pick up the rules of your games!

★ Your dog enjoys playing with toys and balls

★ ★ She runs eagerly after a toy or ball thrown for her

★ ★ ★ She initiates the game, bringing you a toy or ball

Out and About

It is really important these days, with many of us living in urban areas, to be able to train a dog to adapt to life in a city or town. This means being able to take your pet happily out on a leash.

1 It is only safe to take a young puppy out once she has completed her course of vaccinations. This also applies to a mature rescue dog.

2 A young dog will have to cope with lots of strange sights and sounds on the street, so don't be surprised if she seems a little nervous at first in such unfamiliar surroundings.

3 Go out relatively early in the morning—perhaps at a weekend— when things are quieter, so your dog can calmly get used to the experience of walking along sidewalks with you.

★ Your dog appears nervous on the streets

★ ★ She tends to stop and sniff repeatedly on a walk

★ ★ ★ She walks alongside you and is responsive

Play an energetic game with a young dog before you venture out. She needs to burn off some excess energy before you start walking round the block together.

Rules of the road

For the convenience of other pedestrians, your dog shouldn't pull diagonally across a sidewalk. She needs to walk in a straight line. This is where earlier leash training, walking alongside a fence or wall near your home, should prove helpful (see page 46).

This Beagle is clearly very relaxed about walking along a street. She is following her owner's pace nicely, and is not distracted by scents or noises.

Top of the Class?

How is your dog scoring when it comes to mastering the basics? Don't forget: patience is a virtue and don't despair if progress isn't always smooth. Keep practicing, and you'll get there in the end!

Record how your pet is progressing on the special star chart opposite. Enter the number of stars scored in each quiz into the boxes. For "True or False" questions score 3 stars for the correct answer but none for the wrong one.

Add up the total number of stars your dog has scored in this quiz and then turn to the final score chart on pages 186–187. All will be revealed! You'll be able to work out just how smart your pet is.

How Your Dog Learns

Pages 12–13
Superdog or Doggy Dunce?
☐ Number of stars

Pages 14–15
Genes and Genius
☐ Number of stars

Pages 16–17
Fit for Fun
☐ True

Pages 18–19
Man's Best Friend
☐ Number of stars

Pages 20–21
Brain Games
☐ Number of stars

Pages 22–23
School of Fun
☐ False

Pages 24–25
How Dogs Learn
☐ Number of stars

Pages 26–27
Smart Puppy
☐ Number of stars

Pages 28–29
Quick to Learn
☐ Number of stars

Pages 30–31
Old Dog— New Tricks?
☐ Number of stars

Pages 32–33
Mischief Makers
☐ Number of stars

Pages 34–35
Boredom Busters
☐ Number of stars

Pages 36–37
Dog's Eye View
☐ True

Teaching the Basics

Pages 40–41
The Name Game
☐ Number of stars

Pages 42–43
Dog Days and Nights
☐ Number of stars

Pages 44–45
Chilled Out
☐ Number of stars

Pages 46–47
Walkies!
☐ Number of stars

Pages 48–49
His Master's Voice
☐ True

Pages 50–51
Sitting Comfortably?
☐ Number of stars

Pages 52–53
Sit and Stay!
☐ Number of stars

Pages 54–55
Total Recall
☐ Number of stars

Pages 56–57
Fun Times
☐ Number of stars

Pages 58–59
Out and About
☐ Number of stars

How did my dog score?

★ Mostly 1 star = more work needed!

★ ★ Mostly 2 stars = your dog is becoming a brainiac

★ ★ ★ Mostly 3 stars = your dog is a gold star pupil

It's a Dog's World

Clock Watching

Can your dog tell the time? There's good evidence that he can, based on the behavior of many pets. Although it is difficult to prove conclusively, certainly dogs do establish their own routines, just as we do. Their body language shows that they know when meals are due, or walks are coming up.

1 You feed your dog at set times each day. Your pet will soon realize when these are, and if you forget, expect to be reminded with a gentle nudge!

2 Your dog may follow you around more closely than usual, perhaps whine, and he may scratch at or even pick up his food bowl to attract your attention.

3 Should you forget a walk, then your pet will leap up in anticipation when you move, and may even bring his leash to you if he can reach it.

A dog may not recognize the time as such, but his inner clock will soon let him know when to anticipate a meal or a walk.

★ Your dog knows his meal times

★ ★ He knows when it is time to go for a walk

★ ★ ★ He is waiting to greet you when you get home

If you go out and return at a set time each day, then it's very likely that your dog will be waiting at the door to greet you with a wagging tail and a wet tongue.

Home alone

The ability of a dog to know when you are coming back may not necessarily be related to time, as you might first think. Their keen hearing means they can detect the sound of a car engine or the crunching of gravel underfoot, which may well be inaudible to people indoors.

Best of Friends

Dogs, thanks to their social natures, can bond well with other family pets, and sometimes strike up quite unusual friendships. Nevertheless, you need to be very careful, particularly when it comes to introducing small pets because, despite cute videos on the Internet, many dogs, such as hounds and terriers, were originally bred to hunt.

A kitten often appreciates the company of a placid dog when it comes to settling in at a new home, after leaving its mother and littermates.

GETTING ALONG

1 To ensure a harmonious household, feed your pets in separate rooms. Otherwise mealtimes can develop into serious flash points. If trouble does break out over food, it can be a difficult issue to resolve long-term.

2 Do not force your pets together. Allow them to get to know each other over time, and watch them carefully from a distance when they are together. Watch for warning signs that may signal that trouble is brewing.

Cohabitation
If you are searching for a companion to live alongside an existing pet, try to find out if the dog you have in mind has lived with other pets before.

3 Adult cats will adjust to the company of a lively puppy, and tend to be the boss in such situations. If things get a bit too boisterous, the cat will probably hiss and simply walk away.

★ Your dog and cat are happy to be in the same room

★ ★ They recognize and approach each other in the yard

★ ★ ★ Both pets are happy to sleep close to one another

Dogs can distinguish between cats that are part of their "pack," and others wandering through the neighborhood. They may even dive in to see off an interloper if a fight threatens.

Puppies and cats usually become friends over time, but may be wary at first.

Smart Moves

No two dogs are the same. Each dog has his own character. Indeed, it is quite common for two individuals of the same breed, who may possibly even be littermates, to display quite different personalities when they are sharing a home together with you.

1 Dogs in a group establish a sort of pecking order, with some being more assertive than others. This becomes very apparent when they are living at home with you.

2 Observe the personality of your pet emerging. His early life has a huge impact on how he will develop. Early socialization with people is important to ensure that your puppy grows up confident.

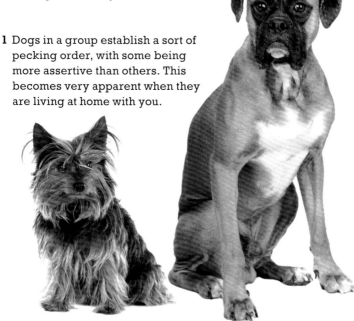

3 It will soon be obvious which out of two dogs is the dominant individual. Their body language when together will show you which one is more assertive by nature.

Don't immediately assume that a bigger dog will be more dominant. This often proves not to be the case, particularly if he is the newcomer to the family. Small dogs are often very feisty.

Above: The collie on the left, with his ears drawn back and lowered, is not challenging his companion. If conflict between males becomes an issue, sometimes neutering lessens the risk of continuing aggression.

Right: A distinct pecking order will develop in a litter of young puppies, and this will remain as the dogs grow up. The Shar Pei that is sitting up on the left is the dominant individual here.

Dogs do differ
The way that dogs react is affected not just by their ancestry, but also by their upbringing. A rescue dog may easily be scared of something seemingly innocuous that recalls an unhappy event in his past.

Clickety Click!

Clicker training is a really simple but effective way of training your dog. The handheld clicker is used to make a distinctive sound during training, which "marks" a specific action that you want. This helps your dog by making clear that he is doing just what is required.

1 Clickers are available online or in pet stores, and are easily incorporated into a training program, whether you are working with a puppy or older dog. Your pet will soon learn the sound and react positively to it.

2 Once your dog displays the required behavior, such as sitting when asked, immediately click. Then reward your pet with a treat. He will soon begin to realize that sitting brings rewards.

By emphasizing what a dog is doing correctly, clicker training is referred to as "positive reinforcement." Some dogs learn to respond very quickly, often within just a few clicks.

The clicker's value

Dogs are very responsive to sounds. While they come to recognize our voices, the beauty of using a clicker is that it is a unique noise. Never use it in other situations for this reason. It is important that the dog only associates the sound of the clicker with his own behavior. If your dog doesn't respond to a clicker, it may be due to deafness.

3 The clicker helps your dog understand what is required, and so he will often concentrate harder, waiting to hear its distinctive sound. This helps to make training more fun for your pet.

★ You click several times but your dog just looks at you

★ ★ He concentrates while waiting for the clicker sound

★ ★ ★ Your dog responds positively to a single click

Above: Don't give a treat every time that you click. If you do, the dog's focus may switch to the food reward, which will undermine the clicker's usefulness.

Left: Clickers come in various designs and do not require batteries. Most can be attached to keyrings. If you lose or break yours, try to get a matching replacement.

Dig It

As everyone who has watched a childrens' cartoon knows, dogs love to scrabble at the dirt with their paws to bury bones in the yard! In fact, not all dogs do act like this, it depends on the individual. But why do they do it?

1 A dog typically starts off by playing and gnawing on a bone, and usually will only attempt to bury it later, carrying it around in his mouth, pausing, and stopping to sniff the ground to find a suitable spot.

Above: Many dogs like to gnaw on a hard bone or chew toy. This helps to prevent tartar building up on their teeth.

Below: Don't offer your dog raw bones. Always cook the bone thoroughly and allow it to cool. Choose thick marrow bones. Don't use poultry bones as they can splinter.

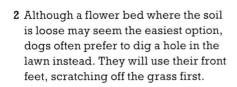

2 Although a flower bed where the soil is loose may seem the easiest option, dogs often prefer to dig a hole in the lawn instead. They will use their front feet, scratching off the grass first.

3 Sometimes they don't do a very good job of hiding their bone, making only a feeble attempt to disguise its presence by scraping back the soil over it. The hole may also not be deep enough to do the job properly.

★ Your dog takes a bone and gnaws it

★ ★ He starts to dig holes when he has a bone

★ ★ ★ Your pet buries his bone in a hole he has dug

Wolfing it!
This behavior probably reflects the way that gray wolves behave in the wild. When they have made a kill and have a surplus of food, they try to cache it, hiding it away from scavengers, often by attempting to bury it.

Left: Rawhide chews are popular today, and may be gnawed away by your dog, with small pieces being swallowed. These are less likely to be buried than solid animal bones.

Below: Dogs can dig easily in soft beach sand, and may try to bury a small ball or toy when playing there.

Goggle Box

Do dogs like to watch television? Some basically ignore it, but quite a few others get actively drawn into the experience, even barking at other dogs or animals that appear on the screen. They can be fascinated watching DVDs, and may even track the progress of the ball when watching sports programs.

Cartoons often hold a dog's attention, and special DVDs are now being produced for dogs to watch while their owners are out. It's a way of preventing boredom that can lead to destructive behavior.

1 Dogs often become interested in the television when they are sitting down with you. Usually most dogs will tend to go to sleep, but keen doggy TV addicts will stay awake, focusing quite carefully on the screen.

2 More enthusiastic participation may lead to your dog jumping up and rushing toward an animal appearing on the screen, barking loudly as it does so. Be careful with large dogs as they may knock the set over.

★ Your dog shows an interest in your TV set

★ ★ He actively responds to what is being shown on screen

★ ★ ★ He has his own dog DVD to watch at home

Why the interest?

In the past dogs usually ignored television, but now, thanks to the higher resolution of today's sets, it is easier for them to resolve a clear image. But they don't see the picture in exactly the same way as we do, partly because their color vision is not as acute as ours.

Above left: If your dog starts to bark during a program, you may need to change channels, or squeeze an empty plastic bottle to create a distracting sound.

Above right: Fortunately dogs don't actually demand to use the remore control! But they do show a preference for certain types of program. Animated cartoons are often a favorite.

On the Move

Just like we humans, dogs can suffer from travel sickness when they first start traveling by car. Take precautions, ideally by not feeding your pet just before you go out. Pack a cleanup kit with you, and make sure that your dog is safely confined in the back.

STARTING OFF

1 With a young puppy, begin by taking your pet out for short journeys on a regular basis to get him used to the unfamiliar sensations of traveling in a car in strange surroundings.

Left: Placid, well-trained dogs can wear a canine seat belt. A seat cover is recommended to protect the upholstery.

Above: Dog crates keep a dog safe in the trunk of the car. Get your pet used to his crate at home before you transfer it to the vehicle.

2 Don't only take your dog out in the car for walks. Otherwise, he will become conditioned to think this will always happen, and he may start misbehaving by barking if he doesn't get a walk.

3 If your dog has developed a fear of car travel because of car sickness, let him get used to being in a stationary vehicle. You can even give him meals in the car to help with this phobia.

★ Your dog is excited about going out in the car

★ ★ He likes getting in the car but doesn't bark

★ ★ ★ He settles down quietly to enjoy the ride

Below: Never, ever leave your pet on his own in a car. The temperature inside can rise to a fatal level in minutes. Always travel with a bottle of water and a bowl.

Above: Sitting pretty. Dogs can enjoy more unconventional rides—but make sure your pet is safely secured so that he cannot cause an accident.

Safe journeys

Always remember to keep your dog safely confined, in a pen (crate) in the back of the car, or behind a dog guard. Special seat belts for dogs are available, but these keep the dog closely confined, and so can cause your pet to become restless.

Out in the Open

Dogs love to explore unfamiliar surroundings, and generally they won't come to any harm as they are instinctively wary of anything they do not recognize. Puppies are more likely to end up in tricky situations, simply because of their inquisitive natures. If you feel anxious in any situation, don't hesitate to put your pet back on the leash.

LEADER OF THE PACK

1 Puppies normally learn about the world around them from their mother and other pack members, and you need to fulfill this role in their place. Be alert to possible dangers, especially when out in unfamiliar surroundings.

Left: Unexpected dangers can include snapping at and being stung by a passing bee. If the dog's tongue swells up, breathing becomes difficult. Call the vet!

2 Be sure that your dog responds reliably to the basic instructions of stop, sit, and come. By using these commands you can prevent your pet from careering into danger, perhaps by chasing a rabbit across a road.

Young explorer

Exploring is part of a dog's nature, and it is very important for his development. Puppies in particular need this experience. A dog that grows up largely isolated is likely to be nervous by nature for the rest of his life.

★ Your dog walks with you but may sometimes wander off

★ ★ Your pet stops when you call him

★ ★ ★ He returns immediately to you when called

Sometimes it's hard to see what has attracted your pet's attention. Dogs tend to be instinctively cautious, so this gives you a chance to call your pet back to you if you sense that something isn't right.

3 Always keep an eye on the route ahead when you are out with your dog. Every now and then, reinforce the recall training by calling your dog back to you and making him sit.

Social Circles

Mixing with other dogs is a feature of canine life in parks and the wider world. But just as with people, some dogs get on better together than others, for reasons that are often unclear to us!

1 There is a crucial phase in a puppy's life known as the socialization period, extending from about seven to 16 weeks. This is when your puppy needs to meet other dogs to lay the foundations that will underpin good relationships in the future.

2 Veterinary practices now offer puppy socialization or play classes. Make inquiries about this, because meeting and playing with other dogs early on is very important to help your puppy grow into a well-balanced dog.

★ Your puppy plays well with other puppies

★ ★ Your older dog gets on well with other dogs

★ ★ ★ Your pet has his very own friends

3 Bear in mind that certain types of dog are less social than others. Pack hounds such as Beagles were bred to live in groups, whereas terriers have more solitary natures. Some breeds have dog-fighting ancestries.

Below: Dogs soon get to recognize other individuals that they meet regularly when out walking. They will often bound up excitedly, sniff each other, and then chase around together if they are off the leash.

Take plenty of time!
Allow dogs to meet on their own terms. Conflict is less likely if they are both off the leash. A dog still on the leash can be nervous of an approaching stranger.

Above: Dogs of different sizes can get along fine together. Disputes are most likely to break out over food or the possession of toys.

Friend or Foe?

When it comes to people, dogs will soon recognize and accept friends who visit regularly, but they may still be suspicious of strangers. This can be a good thing up to a point, but you do not want your pet's wariness with unfamiliar people to turn into a problem.

3 Don't be fooled into thinking that it is just large dogs bred for guarding duties that sound intimidating when they bark. Small breeds such as Pugs can sound pretty fierce too.

1 Much comes back to training your dog to behave in the correct way, starting in puppyhood. This helps to teach good behavior.

2 Your dog needs to learn that barking initially is fine, but when told to be quiet, he must stop and then settle down on his bed.

Dogs can be very affectionate toward people whom they know well, particularly if they have not seen them for some time. Even so, discourage your dog from jumping up at them!

Always reassure your dog that everything is well after he has alerted you to the presence of visitors to your home.

Breed differences

Certain breeds, such as Dobermanns, Rottweilers, and the various types of mastiff, were developed specifically for guarding purposes and, even today, such dogs are far less inclined to be friendly to visitors. Training from an early age can help in this respect, but since such breeds tend to be big dogs, it is worth remembering that they can be intimidating. Only experienced handlers should consider owning large dogs of this type.

★ Your dog barks to warn off visitors

★ ★ He quietens down when he is told to

★ ★ ★ He then settles back into his usual routine

Dogs of mastiff stock, such as this magnificent Neopolitan Mastiff, were originally developed as intimidating guardians. Such breeds are much more docile today, but may still display a marked suspicion of strangers.

Doggy Detective

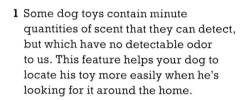

All dogs have a much greater ability to detect smells than we do, and they rely heavily on this sense to learn about the world around them. Those with the best sense of smell are scenthounds. They can be recognized by their relatively long and broad noses, which have a large surface area on the inside of the nostrils to pick up scents.

1 Some dog toys contain minute quantities of scent that they can detect, but which have no detectable odor to us. This feature helps your dog to locate his toy more easily when he's looking for it around the home.

2 When outdoors, dogs like to sniff in prominent positions, such as on lampposts and the base of trees in parks, where other dogs have previously urinated. The scents enable them to determine which other dogs have also left their "calling cards."

3 Young male puppies will squat initially, like bitches, when they urinate. Once they are mature, then they cock their leg as this allows them to spray urine more easily on surfaces where it will leave a distinctive scent.

A dog's nose usually appears slightly moist. The damp mucous membrane is very effective at picking up chemical molecules in the air. A dry nose, however, is not necessarily a sign of illness, as is often believed.

Jacobson's organ

Located in the roof of the mouth, the Jacobson's organ allows dogs to detect pheromones—chemical messengers that are wafted through the air. This method of scent detection explains why male dogs can easily track down a female in heat. Never take a bitch in heat out for a walk off-leash for this reason.

True or False?

Scent-detecting cells in the dog's nose cover an area 50 times larger than in our human noses.

☐ True
☐ False

Below: The Bloodhound is a very effective "cold tracker," being able to follow scents over long distances that may have been laid down several days beforehand.

Right: In the United States, coonhounds were developed for tracking raccoons that climb trees, rather than remaining down on the ground.

Deputy Dog

Dogs are routinely used in the field of law enforcement in various ways. Perhaps most conspicuously, dogs are seen working in public places such as airports and docks, where they are trained to detect explosives and illegal drugs by smell. The dogs used are accustomed to working closely on a one-to-one basis with people, and they are frequently gundogs such as as spaniels and Labrador Retrievers.

German and Belgian Shepherd Dogs are preferred for more direct law enforcement tasks, like chasing and apprehending criminal suspects. Their courage, pace, strength, and intelligence are highly valued by their handlers.

A close bond develops between handler and dog. Their lives may depend on the strength of their working relationship, and how well the dog has been trained.

Left: Dogs' keen sense of smell allows them to detect illegal substances concealed in luggage and parcels. Dogs are trained to recognize specific scents, and they can even pick them up on a person's clothing.

Below: Training of police dogs has to be very rigorous. It helps if the dog can intimidate a suspect, but the dogs must learn not to attack and injure someone when apprehending them.

Did you know?

The use of dogs in law enforcement stretches back to the Middle Ages, when Bloodhounds were employed to hunt down criminals. Bloodhounds are still used today to hunt for people who have gone missing in areas which would be difficult to search by other means.

Sounding the Alarm

As well as superior scenting skills, dogs also possess much better hearing than us, and this is why they are highly valued as guard dogs, particularly during the hours of darkness. These acute nighttime senses allow them to detect intruders very effectively when we are most vulnerable.

1 Puppies are not as vocal as older dogs, and generally have a higher pitched bark. Even so, they will bark quite persistently if they become aware of something unexpected.

2 While having a dog that barks can be useful on occasions, you do not want your pet to bark persistently. Some breeds, notably sighthounds such as Whippets and Greyhounds, are relatively quiet by nature.

3 A dog that barks persistently is very annoying and may lead to complaints from neighbors. The worst situation is if your dog barks persistently when you go out, so check with your neighbors if this is a problem.

Dogs do not just bark to warn of a possible intruder. They may bark to express excitement, or indeed frustration, if they feel they are being ignored. The tone of the bark will help to indicate why they are making such a noise.

Tone it down!

If you want your dog to stop barking, don't be tempted to shout in the hope of persuading it to quieten down. This will probably make the situation worse, as your pet will think that you are joining in too, as happens with a pack of dogs. Instead, speak calmly in your normal voice.

Below: You can usually recognize when something is wrong when your dog barks unexpectedly at night, especially if this is accompanied by growling. Old dogs sometimes suffer nighttime anxiety though, which can also cause them to behave in this way.

★ Your dog barks when he is excited

★ ★ He barks to warn you if he hears a stranger

★ ★ ★ Your pet stops barking when he is told to

Far From Home

There are many remarkable stories of dogs being lost and yet managing to find their way home over considerable distances, even from places which are unfamiliar to them. For instance, in 1979 a female German Shepherd dog called Nick was lost on a family vacation in Arizona. Four months later she turned up back at home in Washington— nearly 2,000 miles away!

1 Obviously, the aim is not to get into a situation where you become separated from your pet. This is why it is so vital to be able to call your dog back consistently when you are out for a walk. Keep practicing this recall to reinforce the training process.

Your dog is most likely to get lost in an unfamiliar area. Don't always assume your dog has run off though, because terriers like this Jack Russell may delve into underground burrows.

2 Unfortunately, some dogs, particularly scenthounds such as Beagles, will pick up scents and dash after them, apparently deaf to your calls. The key thing is not to chase after your dog, because he will think this is all part of the fun. Stay where you are and call and whistle continually. Normally he will return when the trail goes cold.

★ Your dog responds when you call his name

★ ★ He will stop running from you when you ask him to

★ ★ ★ Your pet returns to you when he is called

Chipper!
Be sure to have your pet microchipped as a puppy. Then if he is lost and subsequently found, it is easy for you to be reunited, as the chip bears your personal contact details.

Stray dogs can survive by scavenging, but they are at particular risk of being run over when trying to cross roads. Dogs generally display little road sense when out on their own.

A Dog's View

So how is your dog doing when it comes to assessing his place in the world, and how he relates to his environment? Having a well-adjusted dog helps to ensure that you too can have a more relaxed lifestyle.

Record how your pet is progressing on the special star chart opposite. Enter the number of stars scored in each quiz into the boxes.

For "True or False" questions score three stars for the correct answer but none for the wrong one.

Add up the total number of stars your dog has scored in this quiz and then turn to the final score chart on pages 186–187. All will be revealed! You'll be able to work out just how smart your pet is.

A well-socialized dog will get on fine with other animals and with your family and friends. This should be a goal of your training exercises..

It's a Dog's World

Pages 64–65
Clock Watching
☐ Number of stars

Pages 66–67
Best of Friends
☐ Number of stars

Pages 68–69
Smart Moves
☐ True

Pages 70–71
Clickety Click!
☐ Number of stars

Pages 72–73
Dig It
☐ Number of stars

Pages 74–75
Goggle Box
☐ Number of stars

Pages 76–77
On the Move
☐ Number of stars

Pages 78–79
Out in the Open
☐ Number of stars

Pages 80–81
Social Circles
☐ Number of stars

Pages 82–83
Friend or Foe?
☐ Number of stars

Pages 84–85
Doggy Detective
☐ True

Pages 88–89
Sounding the Alarm
☐ Number of stars

Pages 90–91
Far From Home
☐ Number of stars

Dogs of different breeds can get along very well, especially if they have grown up together.

How did my dog score?

★ Mostly 1 star = more work needed!

★ ★ Mostly 2 stars = your dog is becoming a brainiac

★ ★ ★ Mostly 3 stars = your dog is a gold star pupil

First Fun Tricks

Saying Howdy!

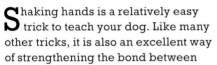

Shaking hands is a relatively easy trick to teach your dog. Like many other tricks, it is also an excellent way of strengthening the bond between the two of you. This one can even help you to improve her well-being in certain situations, such as when removing thorns from her pads or clipping her claws.

LET'S SHAKE ON IT

1 Begin with the basics. The first vital step is to teach your dog to sit on command, and remain in this position (see pages 50–51).

2 Then start stroking your dog while kneeling in front of her, and gently lift one of her front paws a little way off the ground.

3 Do this with both front paws in turn, but only for a few seconds. Over time, lift each leg slightly higher, eventually up to the level of her chin.

Initially replace your dog's foot on the floor after she has shaken hands, rather than leaving it in midair. In this way your pet will start to learn the routine.

★ Your dog sits still on command

★ ★ She raises her paw when you say the word "paw"

★ ★ ★ She places her paw in your outstretched hand

4 Say the word "paw" just beforehand, so your dog learns to recognize a verbal cue for what is expected. Before long, she will raise her paw for you when you ask her to.

Clipping your pet's claws will be much easier if she allows you to touch her feet, as shown by this well-pedicured English Bulldog.

Sit and Beg

Sitting and begging usually proves to be quite an easy combination of movements for dogs to learn. Sometimes your pet may start to behave in this way almost instinctively, because dogs often realize that it is an effective way to attract your attention.

THE ART OF PERSUASION

1 Once your dog is sitting, you must persuade her to look up by holding a treat above her head, just out of reach. Your dog will want to investigate what is on offer.

2 She is likely to move her head up in the direction of the treat. When she starts to move, lift your hand slightly, so as to encourage your dog to raise her forelegs up off the ground.

It is a good idea to start teaching your dog how to sit and beg after an energetic walk, so she will be less inclined simply to jump up.

Natural behavior

Sitting is a natural posture for dogs, as we can see from the behavior of the gray wolf, which is their direct ancestor. Wild dogs in general are also able to support themselves on their hind legs when they need to, which explains why domestic dogs are able to beg.

3 Dogs may learn to beg spontaneously, lifting themselves up in this way to get your attention. This is most likely when you are sitting on a chair, and they feel they are being ignored. Begging on their hind legs will get your attention.

The offer of a tidbit helps to encourage your dog to sit up and beg, but give a specific verbal instruction too, so your pet learns what is expected.

Apart from using their hind legs to support themselves, dogs also rely on their tail which helps to provide extra support.

★ Your dog sits readily on command

★ ★ She starts to sit upright when offered a treat

★ ★ ★ Your pet masters the ability to beg when asked

High Five!

Why not teach your dog to do a high five? If you have already taught her to shake hands (see pages 96–97), you are well on the way to success with this smart trick.

1 Choose a quiet area to work where you and your dog can train together without distractions.

2 Place a small dog treat in your palm, make a fist, and put your hand just under your dog's nose. Hold it fairly close to her face so that she can smell the treat. She will paw at your hand. Don't kneel too far away from your dog or she may lose her balance when she reaches out to touch your hand.

3 Reward your dog with a small treat each time she paws at your hand. Repeat the training until your dog consistently touches your hand.

Treats and tidbits are very useful aids to help you teach your dog specific actions and behaviors.

Paws for thought

Teach your dog to raise her paw only when you ask her to, by ignoring her when she tries to paw for attention on other occasions.

4 Move your hand slightly to either side of her head, so that your dog has to lean over to follow the tidbit, which will encourage her to touch your hand in order to receive the treat.

5 When your dog consistently touches your hand, repeat the exercise without holding a treat. Open your fist and hold your palm vertically in the air. Acknowledge each successful touch with a treat.

6 Add a verbal cue, by saying "high five!" each time your dog touches your palm with her paw. Soon your dog will learn to high five on your voice instruction alone.

★ Your dog touches your fist when holding a treat

★ ★ She touches your open palm without a treat

★ ★ ★ Your dog can "high five!" on voice command

With a bit of patient training and a few small dog treats, you should soon have your pet giving you a cool high five!

Salute the Flag!

Impress your friends by getting your dog to salute on command! It is relatively easy to teach your dog this trick as it is simply a variation on the basic high five (see pages 100–101), although some dogs do find it slightly harder to master. Also some dogs find saluting more challenging than others because of their physical shape.

It is easier for terriers, which have short, fairly compact bodies, than it is for slender, long-bodied breeds such as Greyhounds which have difficulty in sitting comfortably for long.

1 The aim of this trick is to get your dog to sit still, raise her paw, and then keep it still for a few moments, close to the side of her nose in a manner resembling a miltary salute.

2 The starting point is the same as for high five. Teach your dog to sit, so she is relaxed, and go from there.

Your dog will not be able to swivel her paw right round to touch her head, but she should still be able to perform a convincing salute.

3 If you have taught high-fiving first, your dog will be used to raising her paw readily on command.

4 You need to pull your hand away from the high five, without letting your dog put her paw back on the ground. Add the verbal instruction "salute" as your dog performs this cute trick.

Don't let your dog become too focused on a treat when training. You want her to concentrate on what you are saying.

> ### Eyes front, paw raise!
> The flexibility of a dog's paw is less than that of the human wrist and hand, and so a dog will instinctively keep her paw parallel to her face.

★ Your dog will sit still when asked to "sit"

★ ★ Your dog can raise a paw when asked

★ ★ ★ She will salute and hold it for a moment

Roll Over

Your dog is often likely to roll over, as this is part of her natural behavior when she is relaxed. However, the trick is to get her to roll over on command. The gender of your pet can make a difference here though, as bitches are more inclined to roll over instinctively than male dogs.

1 Since this is a posture readily adopted by dogs, simply make a fuss of your pet while she is sitting on the ground, and before long she should roll over.

2 Dogs like having their tummies tickled and stroked. This behavior reinforces the bond between you, as it shows that your pet trusts you.

Rolling over is something that dogs will sometimes do of their own accord, wriggling and stretching their back muscles.

3 When your dog is lying on her side, move her gently so that she ends up lying on her back, adding an instruction such as "over," which she will soon recognize and learn to obey.

4 This is one trick where treats are really redundant to the learning process, partly because your dog ends up lying on her back! Stroking and praise are suitable rewards for this performance.

★ Your dog rolls onto her back when you encourage her

★ ★ She will roll over when you ask her to

★ ★ ★ Your pet will get up again on command

Train your dog to adopt this position by gesturing with your hand in a rolling motion, and then encourage her to stay by using the appropriate hand signal.

Up and down!
After greeting you excitedly, your dog is quite likely to roll over instinctively, hoping to be stroked. Persuading your pet to get up again may be harder!

Playing Dead

Let your dog play for several minutes to burn off some energy before you start training. Then give your dog a few moments to relax and cool down. This should stop her from getting distracted while training how to play dead.

Training should always be fun for your dog. If you see that she is getting distracted, or frustrated, give her a short break or postpone the lesson until later in the day.

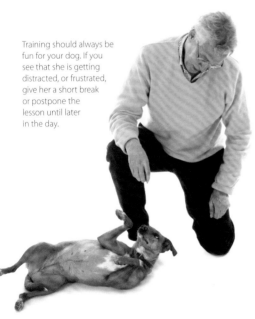

1 Pick a flat area in a peaceful part of the house and start by getting your dog to sit and then lie down as the starting point. You are building on your previous training sessions.

2 Begin training by saying "bang," while pointing your hand at her in the shape of a toy gun. You are trying to use normal verbal instructions to get your dog to sit, lie down, and then, if she is familiar with this, roll over.

3 Gently use your hand to roll your dog onto her back and to hold her still without flopping over completely. Praise your dog and give her a treat. You may need to practice this several times until your dog understands how she is expected to respond.

4 Repeat steps 2 and 3 until your dog has mastered them. It is better if you help her to get these separate movements locked in her mind before you ask her to do them on her own.

5 Now try to condense the verbal instructions (sit, down, roll over), by saying "bang" while using your hand/gun signal to encourage your dog to play dead.

6 If your dog tries to roll over completely every time, stop her legs in midroll by gently putting your hand on her stomach. This helps to teach her that you do not want her to roll all the way over onto her side.

Keep it short
For best results, keep your dog focused on you all the time you are training her. This may require short learning sessions, with plenty of praise, and not too many dog treats.

BANG!

Flat out! Some dogs really get into the spirit of this game. They love to play dead with you.

★ Your dog rolls onto her back . . . and over

★ ★ She follows all your verbal commands

★ ★ ★ "Bang" and your dog plays dead

Three Pot Trick

Test your pooch's powers of observation, memory, and sense of smell with this simple game. It's fun to see just how smart your pet can be when confronted with this challenge.

Left: Dogs have a much more sensitive sense of smell than us, so move all the pots each time. Otherwise, your pet may be confused by the scent of a previous treat.

Above: At first your dog may dive under the container to find the treat. Your goal is to train her to point instead.

1 You need three solid containers, such as small tubs or buckets, under which you can hide a tidbit. You also need some tasty treats for your pet that can be used in training.

2 Now set the containers out in a row on the floor, and place a treat under each of the pots. The promise of something nice to eat will draw your dog's attention to them.

3 Once your dog realizes what the game's about, change the routine. Place a treat under just one container, without letting your pet see which it is. She's got to be smart to find it.

★ Your dog turns over the pots looking for a treat

★ ★ Your pet finds the treat under one of the pots

★ ★ ★ She indicates which pot the treat is under

It's a real thrill when your dog learns to point to the container under which the treat is hidden.

True or False?
Bloodhounds have been known to follow a scent trail for more than 125 miles.

☐ True

☐ False

Hide and Seek

This is a great game to play with your dog when it is too cold or wet to spend long outdoors. It also helps to satisfy your pet's natural curiosity. She must rely on both her scenting skills and natural intelligence to find the toy that you've hidden.

1 Choose a toy that your dog has played with before, so she will recognize its scent. This will allow her to hone in accurately on it when it is hidden somewhere in the room.

Initially let your dog watch you hide the toy, so that she will begin to realize that she must find it so that the game can continue.

2 Allow your dog to sniff the toy first, and perhaps play with it for a few minutes before you hide it. This should help to concentrate your pet's attention on finding that particular item.

3 Make it relatively easy at first, placing the toy perhaps in a cardboard box with a loose lid in the center of the room. Once your dog has got the hang of the trick, try adding several more boxes so that the search becomes quite a bit more challenging.

★ Your dog recognizes a particular toy

★ ★ She follows your directions to find that toy

★ ★ ★ She finds it when it is hidden in different places

Damage limitation

Obviously you don't want to encourage behavior that you might later regret, so avoid concealing toys in drawers. Otherwise your pet may scratch at the furniture in an attempt to find a hidden toy, and may easily do so again in the future, even when you are not at home.

Above: Some dogs, such as Jack Russell terriers, love this game and are generally very good at it. It draws on their natural instincts as they were originally bred to hunt rodents in hidden corners.

Right: Dogs usually have their personal favorites when it comes to toys. Try to use a favorite when you play this game so that your dog will be really motivated to find it.

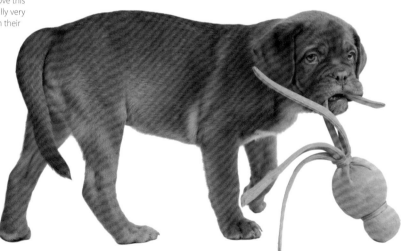

Pass the Toy

It is important that your dog learns from an early age to give up toys and objects when asked. But you don't want this to turn into a drama. Instead, make a game of it, and ensure that you reward your dog with plenty of praise.

1 If your dog has a toy in her mouth, she must learn that if she wants to go on playing the game, then she must give the toy back to you.

2 Give the command "drop" and, if necessary, gently take the toy from your dog by opening her mouth. This can be accomplished more easily with a puppy, who will soon learn to let go. Teaching your dog to drop on command is a vital lesson. It could be a lifesaver if she should ever pick up something dangerous or toxic.

3 Throw the toy for your pet to chase after and bring back to you so that you can throw it again. You are building up a simple training routine that she will enjoy.

If you want to teach your dog to drop a toy, simply place your left hand (if you are right-handed) over her upper jaw, behind the nose. Gently prize open the lower jaw with the fingers of your other hand so the toy falls out of her mouth.

Fast learners

The ancestry of your dog is a significant factor. Retrievers are well-suited to master this type of skill rapidly and instinctively, because over generations as working dogs they have been bred to retrieve game and then bring it back and give it to their handlers.

Below: Keep the message simple, so your dog knows what is expected. Later you can make the game a bit more complicated.

Right: Dogs love to chase after a toy and bring it back for you to throw again. But it's important that your pet drops it on command.

★ Your dog willingly picks up a toy

★ ★ She doesn't run away and lets you take the toy back

★ ★ ★ She will pass the toy to you when you ask for it

Lending a Paw

Dogs are now being trained to help us in a ever-wider range of activities, providing both social and medical assistance. Most people will immediately think of the guide dogs that are trained to help people with seriously impaired vision, guiding them safely down streets and across roads. Less well-known, perhaps, are hearing dogs which fulfill a similar function for people who are deaf and who struggle to hear the phone ring or the doorbell sound. Dogs can even warn epileptics in advance if a seizure is imminent, so they can take their medication to ward off the threat of an epileptic fit.

Certain breeds of dog are particularly suited to the guiding role. Golden and Labrador Retrievers, as well as German Shepherd Dogs, are most commonly selected for such tasks.

Left: The presence of a dog can give much-needed comfort and reassurance, and specially chosen dogs attend hospitals with their owners for this reason. These friendly therapy dogs can be drawn from across a wide range of breeds or may be simply mutts.

Below: Research has shown that dogs can actively help to encourage children to read. They are far more inclined to read a story aloud to a pet than to a classmate or a teacher.

> **Did you know?**
> There are also schemes that take dogs to visit older people living at home or in residential care. They may have kept dogs previously and really appreciate their companionship, but have become too frail to look after one themselves.

Fetch the Remote

Teaching your dog to fetch the TV remote is very similar to training her to do any sort of fairly complex activity. The process involves repetition of the desired action, rewarding good performance, and not rewarding when your dog doesn't respond correctly.

3 Move a little way away and ask your dog to "fetch." If your dog is already familiar with the command, she will pick the remote up and bring it to you. Praise her and reward her with a treat.

Dogs learn mainly by familiarization with actions that are repeated, especially if reinforced with praise and small treats.

1 Remember the steps you have already learned to get your dog to fetch an object. Find an old television remote control that looks like the real one, but which doesn't matter if it gets damaged.

2 Get your dog to sit in front of you and watch as you hold the old remote and slowly place it in the middle of an uncluttered surface.

4 Repeat the training, adding the word "remote" to your command so that she begins to associate the keywords "fetch remote" with this action.

5 Now try leaving the remote in different places and ask your dog to "fetch remote" when you are sitting down. Also extend the distance between you and the remote.

6 With plenty of repetition and rewards, eventually you should be able to sit anywhere you like and simply use the keywords to get your pet to go and find the remote and bring it to you.

★ Your dog finds the TV remote when asked

★ ★ She finds the remote and picks it up

★ ★ ★ Your dog brings it to your seat

Easy does it!
You will be able to train your dog to find and fetch a variety of other objects by using clear keywords that she will learn to recognize.

You need patience to teach your dog as progress may be slow at first, but take your time and always be gentle.

Read All About It

Dogs have long been trained to carry newspapers. Often they combine the task of bringing the paper home with their morning walk with their owner. As newspapers have grown thicker with more sections and color magazines, however, you may have to share the load of a bulky paper between you.

1 The key thing about a newspaper is that it needs to be curled up into a tight roll for your pet to carry easily. An unfolded paper will simply end up being dragged along the ground.

2 Present the newspaper in a rolled-up form to your dog, and she should then carry it home rather like a stick that is picked up when you are out walking together.

A relatively large and soft mouth is to be preferred when a dog is carrying a newspaper, as is the case with this Golden Retriever. The paper is supported behind her canine teeth and, other than a damp patch or two, the newspaper will be fine.

3 Obviously, allow for some setbacks and drops at first, but remember that you can easily teach the necessary skills at home with an old newspaper first to get your dog used to the routine.

It is a good idea to wrap the newspaper in plain white paper, so none of the printers' ink stains your dog's tongue.

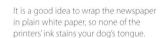

★ Your dog carries toys in her mouth without chewing them

★ ★ She will hold a rolled newspaper in her mouth

★ ★ ★ She regularly carries the newspaper home

The weather

Don't forget that your dog may find it difficult to carry your newspaper on very warm days, when higher temperatures are likely to cause her to get hot and pant. This, in turn, may cause damp patches to appear on the paper.

Early Scores

Here's your chance to see how your pet is progressing overall in terms of mastering these first tricks, although don't be too despondent if the scores are not as high as you'd like! More practice will surely bring success.

Record how your pet is progressing on the special star chart opposite. Enter the number of stars scored in each quiz into the boxes. For "True or False" questions score three stars for the right answer but none for a wrong one.

Add up the total number of stars your dog has scored in this quiz and then turn to the final score chart on pages 186–187. All will be revealed! You'll be able to work out just how smart your pet is.

If you are patient and spend time working with your dog, you will reap the rewards. In a few weeks, you should make great progress.

First Fun Tricks

Pages 96–97
Saying Howdy!
☐ Number of stars

Pages 104–105
Roll Over
☐ Number of stars

Pages 112–113
Pass the Toy
☐ Number of stars

Pages 118–119
Read All About It
☐ Number of stars

Pages 98–99
Sit and Beg
☐ Number of stars

Pages 106–107
Playing Dead
☐ Number of stars

Pages 116–117
Fetch the Remote
☐ Number of stars

Pages 100–101
High Five!
☐ Number of stars

Pages 108–109
Three Pot Trick
☐ True

Pages 102–103
Salute the Flag!
☐ Number of stars

Pages 110–111
Hide and Seek
☐ Number of stars

How did my dog score?

★ Mostly 1 star = more work needed!

★ ★ Mostly 2 stars = your dog is becoming a brainiac

★ ★ ★ Mostly 3 stars = your dog is a gold star pupil

Don't forget to congratulate your pet when the scores are good. Plenty of praise and encouragement is always the best way forward. Keep training sessions short to maintain your dog's focus.

Fitness and Fun

Game On!

Play is a very important part of a dog's life, helping him not just to stay fit, but also mentally alert. It guards against boredom, and helps you to form a closer bond by spending fun time together. As dogs get older, they still enjoy play, but probably less energetically than in the past.

1 Dogs often have a favorite toy, but you may have to replace chew-type toys regularly, especially if your pet is teething, as he can chew them up quite quickly.

2 Different types of dogs prefer different games. Retrievers are very good at finding balls and bringing them back to you, while larger terriers often like to play with tug-toys.

3 Teach your dog to drop his toy or give it back to you willingly. This will help to prevent him becoming possessive about his toys, which can turn into a serious problem.

4 Only play with your pet in a park or other public space when there are no other dogs around. Otherwise, conflicts could break out as other pets want to join in the game.

Some dogs like to play hide-and-seek with soft toys. Make sure that such playthings are in a good state of repair, as loose pieces can be swallowed inadvertently.

★ Your dog will play if you initiate a game

★ ★ Your pet brings his toy back to you when you throw it

★ ★ ★ He brings a toy to you to start a game

Safety matters
Be sure to choose toys that are safe and are specifically designed for dogs. A small ball could prove to be particularly dangerous as it can get stuck in your pet's throat by mistake and choke him.

Above: Balls produced as dog toys are usually relatively soft to guard against possible injury to your pet during play.

Right: If your dog loves to jump and catch things in the air like this, you might want to also get him a special flying disc to play with.

Favorite Toys

Different toys suit different dogs, and their relative size is significant too. What suits a small puppy may not be appropriate for an adult dog, and a ball or any other type of toy that is too big for a small dog to carry easily is likely to be ignored. It will soon become obvious which toy is your pet's favorite, but keep playing with others as well to add variety to your games.

1 Some toys are designed to test your dog's agility, as they bounce at unpredictable angles when they hit the ground, unlike a ball. They are produced in various sizes, so you can find one that will suit your dog ideally.

2 Toys can also improve your pet's dental health, as they are designed for chewing and gnawing. They are more hygienic than bones left lying around the home too. Chewing is a natural part of a dog's behavior.

3 Some toys make a squeaking sound, which attracts the attention of a puppy when you squeeze them. But once he is playing with the toy by himself, the sound often doesn't seem to matter.

Dogs, and particularly older puppies, can become very excited when playing with toys. Tug games are most likely to get them going, so remember to keep the game under control.

True or False?
Larger retailers may now stock more than 300 different types of dog toys.

☐ True
☐ False

Settling-in toys
If you get a puppy that has been used to sleeping with his littermates up to the time you get him, try putting a safe cuddly dog toy in his bed. It should help your new pet to settle down at night.

Puppies need plenty of physical exercise to use up their energy, even though they cannot be taken for long walks at this early stage in life.

Below: Throwing a toy for your dog to bring back to you when walking is great exercise. It means your pet will cover at least twice as much ground as you do.

Having a Ball

Virtually all dogs enjoy playing with a ball. Even if they have not played ball games in puppyhood, they will learn rapidly how to do so when they are older. The size and weight of the ball are significant though and should be chosen to suit your pet's build.

PASSING OR RUNNING?

1 At first, when your puppy gets a ball, he will not appreciate the benefits of sharing! Instead, he is likely to run away with the ball and will be reluctant to give it back.

2 Stand still, and do not chase after the young dog. Call your pup to you, but if he fails to respond, make it clear that you are giving up the game.

3 Soon your dog should bring the ball back to you. Praise him, take the ball, and throw it again. The puppy will learn that for the game to continue, he must retrieve the ball.

★ Your dog is interested in playing with a ball

★ ★ Your pet catches a ball but does not return it

★ ★ ★ Your dog can control the ball with his paws

Conventional soccer balls are not ideal toys for dogs, simply because their sharp teeth are likely to puncture them. A deflated ball means game over!

Right: Try to discourage puppies from using balls as chews. The covering on a tennis ball can be dangerous if swallowed.

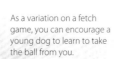

As a variation on a fetch game, you can encourage a young dog to learn to take the ball from you.

Fetch!

Most dogs like to chase after a ball, but the difficult part of this game is to persuade your pet to bring it back to you and drop it. Like all obedience training, you need to teach this in stages. As your pet masters one step, you can then extend the exercise by including another. In the first place, simply choose a ball that is relatively large, so that it is safe and cannot get stuck in your dog's throat. It should also be strong and resistant to the pressure of sharp teeth, but not too heavy.

1 Start by encouraging your dog to recognize the ball and chase after it.

> **Keep your eye on the ball!**
> Always teach your dog retrieve games with a ball on his own, even if you have more than one dog, so that you can secure your pet's undivided attention. Take particular care in public parks, because other dogs may want to butt in and try to steal the ball.

★ Your dog runs after the ball

★ ★ He picks up the ball and runs with it

★ ★ ★ Your dog brings the ball back to you, and drops it

2 Do not roll it too far ahead, so he can see it clearly and run to investigate.

Dogs love to run after a ball and pick it up; the clever part is to train them to bring it back to you and drop it.

3 Most dogs will pick up a ball and then run around with it.

4 Once the puppy has the ball in his mouth, call him to come back to you.

Search and Find

This game is a great way to encourage your dog to use his scenting skills. It gives him a break from chasing after toys while at the same time ensuring that he remains active in both mind and body.

On the right track

A dog's scenting ability varies during the day, being most acute when he is hungry. In addition to finding objects, dogs can track scents too, as in the sport of drag hunting where aniseed is used to create a scent trail.

1 Search and find is a game that you can play with your pet either indoors or out. Firstly get your dog to focus on the toy that you are going to hide. Make sure he sniffs the toy or ball really thoroughly to pick up its scent.

2 In the early stages of the game let your dog see where you hide the toy, so that he learns to go and retrieve it from where it is hidden. Then make it more difficult by getting your pet to search without having seen where you have concealed it.

Get your dog interested in the toy by playing together with it, and let him get a really good idea of its scent before you try search and find.

★ Your dog starts sniffing after a hidden toy

★ ★ He works alongside you looking for the toy

★ ★ ★ He takes the initiative to find the hidden toy

3 This is a game where some guidance from you can be helpful, using hand signals. You need to work systematically alongside your dog pointing to possible hiding places that he can investigate by sniffing.

Above: Scenthounds, such as the Beagle seen here, are well suited to the task of detecting trails thanks to their relatively broad noses which help them to pick up scents.

Left: Spaniels are particularly talented at finding hidden objects, as they were originally bred to find and flush out game that is hidden in woodland areas or in low undergrowth.

Search and Rescue

Dogs are used today in the vital role of search and rescue, saving many human lives every year in disaster zones around the world. Special teams of dogs and handlers are trained for this purpose and remain on constant alert to be flown out to respond to a natural emergency or disaster.

Many types of dog are used, but the St. Bernard is regarded as one of the great pioneers. First kept by monks at a monastery in the Swiss Alps, St. Bernards rescued stranded travelers who had become trapped in heavy snow. They were made famous by the Victorian artist Sir Edwin Landseer. He used a lot of artistic license, however, portraying the dog with a small keg of brandy around his neck to revive people suffering from hypothermia. In reality, this never happened.

Left: Mountain rescue teams still depend heavily on dogs as a means of locating people trapped in avalanches, enabling them to be dug out of the snow quickly. Dogs are sure-footed even when the snow is unstable.

Below: Dogs also work very effectively to locate people trapped under buildings, following earthquakes for example. This is very dangerous work as they must venture into confined areas where the rubble is insecure and could collapse. at any moment

Did you know?

Search and rescue (SAR) dogs fall into two main groups. Trackers are deployed when someone disappears in order to find that particular individual by his or her scent. The other group are air-trackers, which detect the scents of anyone missing in a particular area.

Soccer Star

While most dogs enjoy chasing after a ball, the aim of this game is not to pick up the ball and bring it back to you, but to concentrate on keeping it moving with the dog effectively dribbling with it as he runs. Dogs find this much more difficult, and not all master the skill, so you'll need to be very patient!

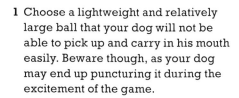

1 Choose a lightweight and relatively large ball that your dog will not be able to pick up and carry in his mouth easily. Beware though, as your dog may end up puncturing it during the excitement of the game.

2 Find a level surface—preferably a tiled or concrete area—where the ball will run around easily. Your pet should then learn to chase the ball around by pawing at it, and so get into the swing of the game.

3 Now, encourage your dog to steer the ball into an area serving as the goal. He will soon come to understand what is required. Finally, teach him to dribble the ball around you and then score a goal. He's better than Beckham!

Even if your dog doesn't turn out to be a soccer star, you can still have fun with him as he learns to control the ball and chase it around the field.

Above: Only kick the ball gently when playing soccer with your dog, and keep it down on the ground to avoid risk of injury to your pet.

Left: Terriers often display good soccer skills as they naturally steer rodents that they are chasing, making them easier to corner and catch.

Groomed to stardom

Although it might seem strange, poodles often have great ball skills. Originally kept for retrieving game from water, which explains their unusual waterproof coat, they were sometimes used as circus performers and as jugglers.

★ Your dog chases after a soccer ball

★ ★ Your pet shows good ball-control skills

★ ★ ★ He is a star player and can score a goal

Flying Discs

Catching a fast-spinning disc in the air requires considerable coordination and agility on the part of your pet, as well as plenty of power in his hind legs to enable him to jump high enough to grab it in his mouth.

1 Start by getting your dog interested in the disc. Throw it gently at a low level over short distances and encourage your dog to run after it and bring it back to you to throw again.

2 Then increase the distance between you and your dog as you throw the disc a little harder. Choose an open area of ground where there is good visibility and no bushes nearby. You may need to practice your throwing technique too!

3 Before long your dog will start jumping to reach the disc as it spins overhead. If the disc falls out of reach, as sometimes happens, your pet can still run to pick it up off the ground.

Staffordshire Bull Terriers love playing with flying discs, and they have the strength and coordination to jump and catch it in midair.

★ Your dog picks up the static disc from the ground

★ ★ He jumps up and catches the disc in the air

★ ★ ★ He catches the disc and brings it back to you

Left: Collies are another breed that has both the agility and intelligence needed to catch flying discs as they whizz past in the air.

No slipped discs!
Always use a special flying disc designed for dogs, rather than a child's toy. These come in different sizes and are generally softer and less likely to cause injury. Furthermore, they are impregnated with a scent that is designed to appeal to your pet.

Even if your dog cannot catch the disc in flight, he will still enjoy chasing after it and retrieving it.

Jogging Partners

Going out jogging or running with your dog is a great way of exercising, both for you and your pet, but you need to choose the right breed for this activity. It must have plenty of stamina and not be liable to keep stopping for a sniff on the route.

BUILD UP STAMINA

1 Just as you would not suddenly begin by running a marathon, don't start by taking your dog out for a long run. Build up gradually over several weeks and monitor your dog's fitness.

2 Be patient and concentrate on leash skills at first. Young dogs must not be overexercised, as this can cause permanent damage to their joints. If in doubt, seek veterinary advice before you start jogging.

3 Avoid running with your pet during the daytime when the weather is at it hottest, as dogs cannot sweat as efficiently as we can and so are at greater risk of suffering heat stroke.

The Dalmatian is a breed that makes an ideal jogging companion. They were originally developed to trot alongside horse-drawn coaches. The breed's conformation and temperament are perfectly suited to even-paced, long-distance running.

Left: Once your dog is well trained, and assuming you are not near traffic, he can be let off the leash and should run happily alongside you while you jog or cycle.

Below: Running along the beach with your dog at sunset—what could be better? If possible, take fresh drinking water and a bowl for your dog, as he will need a drink after exercise.

Footloose and fancy free

Just as you can develop blisters when jogging, check your dog's paws to ensure that they are not sore and painful after running. Dogs can pick up splinters and thorns, and get impacted mud and seeds trapped between their pads.

★ Your dog is keen to run

★ ★ He runs with you but pulls on the leash

★ ★ ★ He runs with you at your pace without pulling

Doggy Paddle

Swimming is great exercise for dogs. Hydrotherapy is now recognized as being a valuable way of aiding recovery from illness or injury, as well as helping those with chronic health issues such as arthritis. Some dogs were bred to work in water, usually as retrievers or to help draw in fishing nets, but dogs in general can swim surprisingly well.

TAKING THE PLUNGE

1 Young puppies may not be very confident when it first comes to swimming, but they will soon master this action with practice.

2 Dogs paddle with their front legs, but the main thrust when swimming comes from the powerful muscles of their hind legs, just as when they are running. Their tail acts like a rudder to help them steer.

3 If your dog is swimming in the sea, wash his coat afterward in fresh water, otherwise he will lick off salt from the seawater when he is grooming himself after the swim.

Hydrotherapy sessions take the weight off painful joints while allowing muscles to be exercised. Special lifejackets are available to keep dogs afloat and make it easy to lift them in and out of the water.

Water dogs

Some breeds, such as the Newfoundland, have webbed toes, which help them to swim more efficiently. The outer layer of their coat is water-resistant as well, so they will not become water-logged or chilled. They just need to shake themselves dry when they emerge from the water.

★ Your dog enjoys paddling in water

★ ★ Your pet likes to swim alongside you

★ ★ ★ He will enter the water and swim without you

Above: Dogs dive into water, and some even enjoy putting their heads under the surface. You will need to dry the pendulous ears of breeds like spaniels as they don't dry easily on their own.

Left: At the seaside your dog may enjoy charging into the water after a ball and even swimming in the sea, but choose an area where you know there are no strong currents or powerful waves.

Water Baby

Once your dog is confident entering water, he will probably want to start playing there, retrieving toys or balls. Usually dogs that have been bred over generations for this purpose, like Golden and Labrador Retrievers, are the most likely to plunge in after a toy.

1 Start playing in shallow water, getting your dog to chase the ball and bring it back to you to throw again. Encourage him to remain focused on the game so that the ball does not drift away, but be prepared to wade in to retrieve it from the water yourself if necessary.

Below: A Labrador Retriever swims back powerfully with a tennis ball that he has successfully retrieved from the surface of the water.

Right: When a dog emerges from the water, the first thing that he will usually do is to shake his coat dry, before dropping the ball.

2 Throw the ball a bit farther out, so that your dog has to venture into deeper water and pick it up where it is bobbing around on the surface. This takes a bit more skill and energy to do.

3 Once this stage is mastered, train your dog to bring the ball straight back to you, so you can throw it again, as you would when playing on dry land.

Left: Some long-coated dogs are going to need a helping hand to get dry, so stand by to give your pet a vigorous rub with a towel.

Above: Confident retrievers will not just pick up toys bobbing on the surface, but will also venture underwater to grab hold of them.

In the beginning
An ancient group of water dogs preceded the retriever breeds. They were used to retrieve ducks and other game shot with arrows. Some of these old breeds, such as the Irish Water Spaniel, still survive today and can be distinguished by their characteristic curly coats.

★ Your dog will retrieve a toy or ball from the shallows

★ ★ He will retrieve while swimming, when encouraged

★ ★ ★ He will plunge into deeper water to retrieve

Skateboarding

Skateboarding dogs have become a YouTube sensation. Naturally some breeds are better suited to this activity than others because of their physique. Fairly low-slung, short-legged dogs tend to be the best skateboarders.

1 The skateboard should be just a little wider than your pet's stance. Start off with the board upside down, so that it is stable, and encourage your dog to stand on it in this position.

2 When you turn the skateboard over onto its wheels, hold on to it so it will not move at this stage. You want your dog to feel relaxed when standing in position on the board.

3 Once your dog is happy standing on the board, start to move it carefully so he gets used to the motion. With practice your dog will be prepared to go faster and may even start to propel the board along himself.

★ Your dog places his front paws on the skateboard

★ ★ Your pet is happy to stand or sit on the board

★ ★ ★ He stays on the board when you move it along

Once your dog is prepared to stand on the skateboard by himself, encourage him to remain on it while you gently push him along.

Skater dude!

You'll need to work hard with your dog to persuade him to take up skateboarding so be prepared to offer plenty of encouragement and rewards. By standing the skateboard on carpeting, you will be able to dampen the movement initially. Then you can gently push your pet back and forth on the board until he gets used to the unfamiliar sensation of skating.

English Bulldogs are up in the top rank of talented canine skateboarders, thanks to their body shape and strength.

Learning by Example

There are many different sports that you can participate in with your pet. You may want to take up serious competition or just to have a bit of fun, but even if you do not aspire to the top competitive level, just working closely together will be very rewarding. The training and interaction involved will strengthen the bond between you.

1 Successful training relies on mastering the basics effectively. This is why so much emphasis should be placed on being able to control your dog through the fundamental commands.

2 Remember to use hand signals. Dogs themselves rely heavily on nonverbal communication, and this method is particularly useful when instructing from a distance or indicating a route.

3 Always remain positive and praise your dog. Dogs learn at different speeds, so be patient and encouraging if your dog appears slow to grasp what it is you want him to do.

You need to be able to guide your dog over and through the obstacles of an agility course. He will be looking to you for guidance.

Back to the future

One of the reasons that dogs make such great companions is the fact that the gray wolf, their ancestor, lives in packs. As a result, they are naturally inclined to form partnerships and social groupings. They have proved to be very adaptable as well as being adept at learning.

★ Your dog reacts to simple instructions

★ ★ Your pet follows your example and copies what you do

★ ★ ★ He readily carries out the skills he has learned

Focused training helps if
there are certain aspects
of an agility course that are
proving problematic for your
pet to master. Use a hoop,
for example, held low to the
ground initially and then
raise it higher in stages.

Dog Agility

Against the Clock

Canine agility is a very popular activity. It shows dramatically just how smart your dog really is, as well as keeping both of you fit. You can do it in your backyard just for fun, or take part in formal agility competitions.

1 Your dog needs to be able to learn all the different components of a course but, as with as with all such training, break the components down into smaller sections to avoid confusing her with too much detail at once.

2 Set up the different sections of an agility course (as described in this chapter), so your pet can practice the various stages individually. It is surprising how fast she will learn.

3 Your dog needs to go through the various parts of the course in the correct sequence, so give the different components names, such as "tunnel" or "weave," and use clear hand signals to point out the correct route to your pet.

Speed is important, but so is accuracy. You are aiming to complete the agility course without making mistakes, such as missing a jump.

Skill sets

Different types of dog are likely to find certain elements of the course easier to master than others. This can be related in part to their temperament, size, and overall physique.

★ Your dog masters one agility activity

★ ★ She moves confidently from one activity to another

★ ★ ★ Your dog can complete the whole agility course

Above: Tunnels are a part of the agility course that generally suit smaller dogs.

Left: Jumping is something that comes more naturally to bigger, longer-legged dogs.

Tunnel Dash

This can be great fun for both you and your dog. Although ideally it should be taught in the yard, smaller dogs can be trained indoors if you have a big enough room to set up a tunnel. Suitable tunnels in a range of different sizes and lengths are available from suppliers of dog agility equipment.

1 Make sure that the tunnel you are using is reasonably secure and that your dog can see right through it to the other end, as she will then venture into the entrance more readily.

2 Go to the far end of the tunnel and call your pet through it. It helps to have someone at the other end to steer her gently into the entrance.

Many puppies are naturally inquisitive and you can easily get them to explore the inside of a play tunnel. They won't then feel nervous when you start training them to run through a bigger version.

3 The next step is to coax your dog to run round to the start once more and to dive back through the tunnel again. Your pet should soon get the hang of this and will enjoy bounding through the tunnel at high speed.

Right: Terriers love tunnels. They have no fear of confined spaces, instinctively venturing underground after prey without anxiety.

A helping hand
It helps to have two people involved in the tunnel dash, until your dog understands what she's supposed to do. Otherwise, she may just run around the side of the tunnel rather than through it.

Left: On a competition agility course, tunnels usually vary in their configuration and length. They often include bends like this

Weaving

It is incredible how a dog can learn to weave instinctively in and out of a series of closely spaced poles arranged in a line. Although always featured as part of an agility event, weaving has become so popular that competitions are held for this activity alone.

Top dogs can weave their way through the obstacle course at speeds equivalent to five poles per second!

Work for it

Do not automatically give your dog a treat at each element, otherwise she will focus on you rather than on the course. You eventually want the dog to run around the entire course before being rewarded.

Border Collies are smart and willing to please, making them ideal candidates for agility tests.

AGILITY TEST—THE WEAVE

1 Before you start weaving through poles, it helps to get your dog to weave around your legs, following a treat held in your hand.

2 Place 12 canes, which are taller in height than your dog, upright in a line in an area of grass. Each one should be spaced about 24 inches apart.

3 Encourage your pet to weave in and out between the line of canes, until she reaches the end . . . and a tempting reward that you have lured her with.

4 See how long it takes for your dog to master what is required. It may be easier to start with just four poles and progress from there, tracking from side to side to call your dog through.

★ Your dog can weave through your legs

★ ★ She is starting to weave through a few poles

★ ★ ★ Your dog can easily weave through all the poles

This collie knows exactly how to weave through a line of poles quickly and accurately.

Climbing Frames

Encouraging your dog to leave the ground and walk across a raised platform can be a challenge, as dogs are not instinctively keen on climbing. However, slats placed on the surfaces of the inclines help your dog to keep her footing, as she moves up and down the ramps of the climbing frame.

1 Place the climbing frame on a level surface and make sure that it does not wobble, as any movement will be very offputting for your dog. Now you must encourage your pet to climb up one side toward the top.

2 Holding a treat usually tempts your dog to follow your hand up the ramp. Go slowly at first, allowing her to gain in confidence as she gets used to the slope. Pause at the top to give your pet a treat and then lead her gently down the other side.

3 With the A-frame, your dog runs up one side to the top of the "A" and then straight down the other side. The incline is steeper than that of a dogwalk, and your dog needs confidence to step over the top.

There are various types of climbing frame—known in agility circles as "contact obstacles"—with the A-frame being based on the design of a step ladder. The key thing is stablity.

★ Your dog runs up the ramp without jumping off the side

★ ★ She goes over the top of an A-frame obstacle

★ ★ ★ Your pet is happy to use a teeter-totter

Below: The dogwalk consists of a central plank fixed around 4½ feet off the ground which your dog reaches by means of ramps.

Right: An A-frame is made like an inverted V. At the bottom is a painted contact area that the dog must touch as she exits the frame.

Teetering

The teeter-totter (or seesaw) is usually the hardest item for a dog to master. She runs up one side to the top and then needs to balance her weight to tip the platform downward to allow her to come down the other side. She must hit a painted contact zone at the foot of the plank before leaping off the seesaw.

Jump For Joy

Jumping is something that comes naturally to many dogs, particularly those with long legs, but even small dogs are happy to jump over a barrier if it is set low enough to take into account their height. Bear in mind when first training your dog over fences that jumping can be hazardous, especially if your pet is running fast and the landing area is uneven. Take it easy to start with.

1 You can buy inexpensive, lightweight jump training sets (and other agility equipment) online. It is a good idea to choose an adjustable set, even if your dog is quite long in the leg.

2 Working with just one jump to start with, encourage your dog to go back and forth over it. If she is reluctant to do so, place a treat on the opposite side of the barrier as an incentive.

★ You dog is happy jumping in both directions

★ ★ She can clear two jumps in sequence

★ ★ ★ She takes all the jumps on a course in her stride

Even short-legged Dachshunds will jump, but be careful with these dogs as they are vulnerable to back injuries.

Above: As dogs become more confident about jumping, so the height of the barrier can be raised. Always check that the landing area is suitable, so your dog won't slip on wet or frosty ground if a tight turn is immediately called for.

Left: These cross poles offer another variation for dogs that enjoy jumping, as they need to target their jump across the lower central area. More experienced jumpers can tackle two jumps in sequence, as shown here, or the poles of a single jump may be made in a spread that calls for the dog to stretch across the jump.

Tire Jump

This is quite a tricky agility obstacle. The dog must get her timing spot-on to jump through the gap, and so good coordination is absolutely vital. Obviously, the size of the tire (or hoop) must be chosen to allow adequate clearance for the size of the dog.

1 Start by using a relatively large tire or hoop, suspending it quite close to the ground at first. You can raise it progressively as your dog successfully jumps through it.

This type of jump can be a bit daunting at first. Choose a soft bicycle tire with a wide aperture when your dog is first learning to jump through.

2 Be sure to keep the tire relatively taut and straight on its supports, because if it moves around, it will be more difficult for your dog to jump cleanly through what is effectively a ring suspended in midair.

3 Once your dog jumps through the tire happily, start to train her to leap back the other way, rather than always taking off from one particular side. Agility courses differ in design, and no two are absolutely identical.

★ Your dog can hop through a large tire or hoop

★ ★ She can jump through a tire raised off the ground

★ ★ ★ Your pet jumps through a tire from either side

Cleared for takeoff

A good jump requires a suitable area for takeoff. Always ensure that the ground is level, and the surface is firm, allowing your dog to pick up pace as she approaches the jump. Check that the supports are securely fixed.

Above: Even small dogs can enjoy taking part in this type of activity, as shown by this lively terrier. The height of the jump in competitions is calibrated to the size of the dog taking part.

Left: Soaring through a tire jump with ease. The dog's powerful hindquarters provide the muscular launch pad, with the forelegs acting as shock absorbers when she touches down lightly on the other side.

Athletic Scoring

You will soon discover whether your dog is an all-rounder or likely to be more of a specialist in the field of agility. There is no doubt that some individuals are smarter than others in this exciting and enjoyable activity.

Record how your pet is progressing on the special star chart opposite. Enter the number of stars scored in each quiz into the boxes. For "True or False" questions score three stars for the right answer but none for a wrong one.

Add up the total number of stars your dog has scored in this quiz and then turn to the final score chart on pages 186–187. All will be revealed! You'll be able to work out just how smart your pet is.

Top agility dogs are really smart! It is not all about pace, but also needs your pet to work actively with you and focus on the course.

Fitness and Fun

Pages 124–125
Game On!
☐ Number of stars

Pages 126–127
Favorite Toys
☐ True

Pages 128–129
Having a Ball
☐ Number of stars

Pages 130–131
Fetch!
☐ Number of stars

Pages 132–133
Search and Find
☐ Number of stars

Pages 136–137
Soccer Star
☐ Number of stars

Pages 138–139
Flying Discs
☐ Number of stars

Pages 140–141
Jogging Partners
☐ Number of stars

Pages 142–143
Doggy Paddle
☐ Number of stars

Pages 144–145
Water Baby
☐ Number of stars

Pages 146–147
Skateboarding
☐ Number of stars

Pages 148–149
Learning by Example
☐ Number of stars

Dog Agility

Pages 152–153
Against the Clock
☐ Number of stars

Pages 154–155
Tunnel Dash
☐ False. The record is a height of 172.7cm achieved by a greyhound called Cindy in 2006.

Pages 156–157
Weaving
☐ Number of stars

Pages 158–159
Climbing Frames
☐ Number of stars

Pages 160–161
Jump for Joy
☐ Number of stars

Pages 162–163
Tire Jump
☐ Number of stars

How did my dog score?

★ Mostly 1 star = more work needed!

★ ★ Mostly 2 stars = your dog is becoming a brainiac

★ ★ ★ Mostly 3 stars = your dog is a gold star pupil

Win or lose, the beauty of agility is that your pet will benefit from it in terms of overall fitness and stamina, and as handler you'll find it a lot of fun too!

Superdog Sports

Canine Sports

Although dogs are not as widely kept for working purposes today as in the past, there are now lots of superdog sports that you can enjoy with your pet that will unlock that working instinct. They range from flyball to dock jumping, or even sled racing. These sports all require a high level of fitness, obedience, and intelligence and are often played competitively.

1 Be sure to match the chosen sport to your dog's natural abilities. Only in this way can you assess how smart your pet is compared to other dogs.

2 It is vital that your dog has mastered the basics of obedience training and responds to simple commands, so he follows your instructions closely.

3 Taking part in canine sports will help to keep your dog fit and mentally alert. It also makes the owner/dog relationship deeper, which helps the whole training process.

Breeds usually excel in the activities for which they were originally developed, so a Labrador Retriever (as shown) brings back objects readily.

Above: It almost defies belief but it's true—dogs do surf! Obviously if you want to take up this sport, it's a whole lot easier if you live by the coast.

Right: Focus is all-important for canine athletes, just as for human ones! It's a great advantage if your dog has good powers of concentration and stays alert.

★ Your dog is fit and healthy

★ ★ He responds well to you and obeys simple commands

★ ★ ★ Your dog shows aptitude in your chosen sport

Fitness first

Many sports require your dog to run and jump. With an older individual, build up his levels of fitness through games before you begin. If you are playing high-energy sports, allow your pet to do some warm-up exercises first to reduce the risk of injury.

Flyball

You can enjoy teaching your dog to play flyball at home if you have enough space. This high-speed agility game, which involves the dog springing a ball from a launcher, catching, and running with it, is particularly good for dogs that are naturally very energetic.

1 Even small dogs can enjoy flyball. It can be enjoyed simply as a game, but serious training is required for the competitive sport version.

2 Competing with other teams against the clock means mastering skills like rapid turns. A quick turn may save vital seconds on your dog's run.

3 As a handler, you must not release your dog early, nor should he drop the ball on his way back over the jumps.

★ Your dog can easily catch a ball

★ ★ He can use the flyball release and catch a ball

★ ★ ★ He can then bring the ball back to you

Flyball calls for natural speed and agility plus intelligence to master the release mechanism. It's a great all-round test of a dog.

Big dogs can just use a paw to trigger the release of the ball, but smaller dogs may have to be taught to jump on the release mechanism to play this exhilarating game.

Doggy Dancing

This popular activity grew out of what is known as heelwork to music. While dogs still weave around their owners, dance routines employ much more freedom than is seen with heelwork.

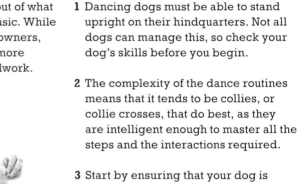

1 Dancing dogs must be able to stand upright on their hindquarters. Not all dogs can manage this, so check your dog's skills before you begin.

2 The complexity of the dance routines means that it tends to be collies, or collie crosses, that do best, as they are intelligent enough to master all the steps and the interactions required.

3 Start by ensuring that your dog is trained to heel, and then you can develop this skill with your chosen music into a dance routine. Clicker training helps too (see pages 70–71).

4 Check out the many YouTube videos to watch a variey of great dancing dogs, which will give you inspiration for your own doggy dances.

Some dogs will readily stand on their hindquarters to look around, or even to steal food off a work surface! You need to develop this skill for successful doggy dancing.

The right moves

Doggy dancing has become very popular, entrancing television audiences and featuring in a movie starring Pudsey, Britain's most famous dancing dog. To reach this standard requires not just lots of practice, but also real talent, and you and your pet must work as a team.

★ Your dog starts to follow you around

★ ★ He has mastered the basics of heelwork

★ ★ ★ He can follow you while standing on his hind legs

Right: Doggy dancing allows your pet's character to shine through in the routine. To get to the top, however, will require massive commitment on your part.

Left: Dressing up your pet can add to the visual spectacle. But the costume must not impede the natural movements of the dog.

Sporting Trials

These provide the opportunity to test a dog's traditional working skills, rather than his appearance, which is normally the focus of attention in the show ring. The spotlight falls on sheepdogs and gundogs, which are graded according to their experience.

1 Gundogs must be able to retrieve target objects from water as well as on dry land. Different breed groups, such as retrievers, are often classed separately, depending on the rules of the trials.

2 Sheepdog trials, also called herding events, involve mainly herding dogs. All entrants must be used to working stock around in the open air, moving sheep into a pen, for example, both quickly and efficiently.

3 These areas are both highly specialized skills, and it is vital to train young dogs in the right way from an early age. Obtaining puppies from a working (rather than a show) bloodline is also advisable.

In normal daily life you generally want to keep dogs away from farm stock, especially sheep, but working sheepdogs will grow up around them.

Right: Gundogs need to be not just physically fit, but responsive too. There are courses available to help you train your gundog, so you can get a good idea of what is involved at one of these.

Below: Encouraging a puppy to bring toys back to you from an early age is a good starting point for training. However, not all retriever puppies will make the grade as gundogs.

Skill sets

Dogs learn to an extent by example, with the skills of working sheepdogs being passed down generations. They even master the ability of controlling stock by fixing them with their eyes. But individual temperament is also important—some gundogs are scared by the noisy guns!

★ Your dog starts to show ability at these skills

★ ★ He responds well to specialist training

★ ★ ★ Your dog is ready to enter events

Dock Jumping

This competitive sport features dogs jumping off a dock or pier into the water. To excel at this activity, you need an athletic dog that loves water, like a retriever. Established rules are laid down for such events, and the top dogs have been recorded as jumping distances of up to 29½ feet.

1 It's a good idea to join a club for this sport. The safety and well-being of the dog are important considerations. A covering to improve grip and avoid slipping is laid down over the dock before dogs start competing.

2 As your dog watches, throw a toy into the water, then lead him back to the starting point, and encourage him to run along the dock and jump into the water after the toy.

3 Subsequently, with your dog sitting at the starting point, throw the toy in from the end of the dock. Your pet should then rush into the water after it when you give him the command.

★ Your dog likes playing with a toy in the water

★ ★ He runs readily along the dock and jumps in

★ ★ ★ He responds to your instructions to jump

You will need a large buoyant lightweight toy to encourage your dog to jump off the end of the dock into the water.

Above: Distance is the key factor by which the jump is judged. So hind-leg strength is very important.

Below: The height of the dock is typically 24 inches, with its overall length being about 36 feet.

Dog Surfing

This amazing sport is something that dogs of various shapes, sizes, and backgrounds can be taught. Obviously, it helps if they have grown up near a beach, so they are familiar with going into the sea from an early age.

SURF'S UP!

1 Start by getting your dog familiar with riding on a board in calm water, and tow him along on it with a piece of cord attached to the board.

2 Once your pet has gained some confidence, you should repeat this while pushing the board over some small waves. With four legs to rely on, dogs can learn to balance well.

3 If your dog is small enough to fit on the board with you, you can paddle around and get your pet used to riding in front of a wave.

Dogs can really enjoy surfing with their owners, but always wash your pet's coat in fresh water afterward to remove any salt that he will otherwise lick off when grooming himself.

★ Your dog swims in the sea

★ ★ He stands on the surfboard for brief periods

★ ★ ★ Your pet is happy surfing the waves

Winning strategies

Competitions are usually held in warmer areas of the world, and they are judged in various ways, such as how long the dog stays on the board, the size of the waves, and how stable the dog is on the board. Dogs can swim well, but beware of currents when he is on the board.

Above: Many dogs enjoy riding on a bodyboard or surfboard on their own, or alongside their owner.

Left: When you are starting to teach your dog, you'll need to steady the board and stay alongside it so that he can gain some confidence before you let him try on his own.

Dog Carting

Dogs have been used to pull carts in various parts of the world for centuries, often providing a vital link to get goods to market. In the frozen north, dogs sleds were until relatively recently the only effective means of transport.

1 Carts come in a variety of designs. It is possible to buy traditional carts by auction, both online and at sales. A cart of this type has two wheels.

2 The cart (or "sulky") is specially designed so it does not place any significant weight on the dog's spine, with the dog being attached to it by a harness. The motive power comes from the dog's powerful legs.

3 "Dryland mushing" is a term used to describe three- or four-wheeled carts pulled by Arctic sled dogs. These are used for training purposes when there is no snow or ice on the ground.

Right: Dryland mushing is also a competitive sport, as seen here. You don't have to use Huskies, other breeds can provide all the power and speed required.

★ Your dog will follow you

★ ★ He will wear a harness without complaint

★ ★ ★ He will walk alongside you pulling a cart

Spreading the load

Only larger, more powerful breeds can generally be used as draft dogs to pull carts. As a guide, carting dogs must weigh more than 33 pounds, and proportionately the total load to be pulled should not exceed three times the weight of the individual dog. Larger dogs can pull a cart with a person onboard. In the 19th century Bernese Mountain Dogs were a popular choice to pull carts with a human passenger. They are still in use in parts of Europe.

Above: Practice carting on level ground, so your dog should not lose his footing.

Below: Swiss mountain dogs have the power and stamina needed for pulling carts.

Mushing

Sled dog racing is a popular pastime, involving a variety of breeds, such as Siberian Huskies that originated in the far North. They run in teams controlled by a sled driver, called the musher, who steers the rig.

GETTING IN GEAR

1 Be aware that mushing can turn into an expensive activity. You are likely to need a team of dogs, as well as the sled, and suitable transport to get them to events.

2 Unlike other canine sports, this does not just rely on one dog, but on a team working together effectively under your leadership. A distinct hierarchy will develop among the dogs.

3 The dominant, most experienced individual needs to be harnessed at the front, acting as the lead dog. This is the most important position in the team, as he sets the pace.

Left: Sled breeds are powerful, lively dogs, which means they need plenty of exercise. Mushing helps to provide them with the active lifestyle they require.

★ Your dog is happy to join a team and fits in well

★ ★ He pulls strongly as part of the team

★ ★ ★ He almost instinctively follows instructions

Steering the course

Sled dogs used to be absolutely vital, moving people and supplies in the Arctic for thousands of years until mechanized transport became available. They also helped to pioneer the exploration of the Antarctic. Competitive racing has developed over recent years, with teams of dogs competing in various formats against the clock.

Above: Training using a cart takes place during the warmer months of the year when there is no snow.

Below: Each dog within the team has a particular role, and they must work in harmony together.

Skijoring

As the word "ski" in the name suggests, this is a sport where the dog pulls a skier along by running over snow. The unusual description derives from the Norwegian word *skikjøring*, which means "ski driving." Powerful breeds are favored for this purpose, and not just sled dogs in this case.

Above: Labrador Retrievers are very responsive when trained, making them ideal for skijoring. They pick up on the skier's instructions very readily.

1 In the first place you need to be a competent skier. Competitions are often held over cross-country courses, but you can also do this just for fun. Naturally you need to live in a state where it habitually snows in winter.

2 Your dog needs to learn to run at a pace with which you are comfortable, and general fitness training is important as well. It is worthwhile having your dog checked by your vet before starting this sport.

3 Be sure that you have the right equipment. Your dog will require a sled-dog harness while you need cross-country skis and a skijoring harness. The two of you are connected together by a length of strong rope.

★ Your dog responds well to walking on a leash

★ ★ He moves as instructed attached to the harness

★ ★ ★ Your dog pulls you along on skis as instructed

Above: Skijoring can involve up to three dogs, and there are ways of unhooking a dog quickly in the event of an accident.

On the move

Events may incorporate both skijoring and sled racing, with the same commands being used in both instances. Sled dogs are therefore able to take part in both types of competition successfully.

Above: Keep an eye on your dog's diet when taking part in cold-climate sports. They burn up calories very rapidly.

An Active Lifestyle

Your dog is unlikely to participate in all the sports described in this chapter, so when you add up the scores only include those events which he has tried. You just want to be sure that he does as well as possible in whatever activities he is taking part in with you.

Basically, these events can be broken down into four areas of activity: agility, flyball, and doggy dancing are the most general categories where all types of dogs can participate. The other three are sporting trials, swimming categories, and winter sports events, which are more specialized.

Even if you do not enter organized competitions, you can put your dog through his paces at home, and see how many stars he deserves!

If you have a purebred dog, you may even enjoy success in the show ring, but he will have to be a very good example of his breed. However, all dogs have some star quality, so be sure to develop the full potential of your pet.

Superdog Sports

Pages 168–169
Canine Sports
☐ Number of stars

Pages 170–171
Flyball
☐ Number of stars

Pages 172–173
Doggy Dancing
☐ Number of stars

Pages 174–175
Sporting Trials
☐ Number of stars

Pages 176–177
Dock Jumping
☐ Number of stars

Pages 178–179
Dog Surfing
☐ Number of stars

Pages 180–181
Dog Carting
☐ Number of stars

Pages 182–183
Mushing
☐ Number of stars

Pages 184–185
Skijoring
☐ Number of stars

Adding Up the Stars

Now it's time to add up the scores of all the stars your dog has achieved in the charts throughout the book. In each case, did they get mostly 1 star, 2 stars or 3?

Page 61
Chart 1
☐ ☐ ☐ Number of stars

Page 93
Chart 2
☐ ☐ ☐ Number of stars

Page 121
Chart 3
☐ ☐ ☐ Number of stars

Page 165
Chart 4
☐ ☐ ☐ Number of stars

Page 187
Chart 5
☐ ☐ ☐ Number of stars

Total
☐ ☐ ☐ Number of stars

How did my dog score?

Mainly 1 star = your dog is smart, but more work needed!

Mainly 2 stars = your dog is becoming a brainiac

Mainly 3 stars = your dog is a gold star pupil

Is my dog improving?

First total star score: Date:
Second total star score: Date:
Third total star score: Date:
Fourth total star score: Date:
Fifth total star score: Date:

Index

Picture credits

Shutterstock.com: 1: Dorottya Mathe; 2*l*: Erik Lam; 2*r*: Africa Studio; 5: Eric Isselee; 6: Sbolotova; 7*al*: Julija Sapic; 7*br*: Maja H; 8*ar*: Jamie Hooper; 8*bl*: Eric Isselee; 9: Ken Hurst; 10-11: Yeko Photo Studio; 12: Annette Shaff; 13: Denis Babenko; 14: luri; 15*bl*: Rita Kochmarjova; 15*r*: Robert Kneschke; 16: Robert Neumann; 17*bl*: Birgit Reitz-Hofmann; 17*r*: Lebedinski Vladislav; 18: Nemanja Glumac; 19*al*: Nikolai Tsvetkov; 19*r*: Erik Lam; 20: Vitaly Titov & Maria Sidelnikova; 21*al*: Africa Studio; 21*r*: Andresr; 22: Anna Hoychuk; 23*al*: Erik Lam; 23*r*: Gelpi JM; 24: Andresr; 25*l*: WilleeCole Photography; 25*ar*: pakornkrit; 26*ar*: Hannamariah; 26*b*: Ermolaev Alexander; 27*ar*: Lex-art; 27*bl*: Damien Richard; 28: PCHT; 29*a*: WitthayaP; 29*b*: Eric Isselee; 30: Ivonne Wierink; 31*al*: siamionau pavel; 31*br*: Simone van den Berg; 32*al*: Wanchai Orsuk; 32*b*: Susan Schmitz; 33*a*: Eric Isselee; 33*br*: AnetaPics; 34: Eric Isselee; 35*a*: Jeroen van den Broek; 35*b*: Eric Isselee; 36*l*: Hannamariah; 36*r*: Dragon Images; 37: Tanyastock; 38-39: Ljupco Smokovski; 40: Dancestrokes; 41: Andresr; 42: mariait; 43*al*: Kalmatsuy; 43*ac*: Sergey Melnikov; 43*b*: Karramba Production; 44: Andresr; 45*a*: Anna Hoychuk; 45*b*: Anna Hoychuk; 46*ar*: Andrey Eremin; 46*b*: cynoclub; 47: Andresr; 48: AnnaIA; 49*l*: Yeko Photo Studio; 49*r*: WilleeCole Photography; 50: Ignite Lab; 51*a*: HelenaQueen; 51*b*: Crystal Kirk; 52: Erik Lam; 53*l*: Jari Hindstroem; 53*r*: Jari Hindstroem; 54: iko; 55*l*: Mikkel Bigandt; 55*r*: Mikkel Bigandt; 56: Yeko Photo Studio; 57: Andresr; 58: George Dolgikh; 59: Monkey Business Images; 60: cynoclub; 62: Eric Isselee; 63: Susan Schmitz; 64*cl*: Dimedrol68; 64*b*: Monika Wisniewska; 65: Anna Hoychuk; 66: Rita Kochmarjova; 67*ar*: Ermolaev Alexander; 67*b*: WilleeCole Photography; 68: Gelpi JM; 69*a*: Eric Isselee; 69*b*: WilleeCole Photography; 70: Barna Tanko; 71*a*: Duncan Andison; 71*c*: WilleeCole Photography; 71*b*: manfredxy; 72*a*: Simone van den Berg; 72*b*: kudrashka-a; 73*l*: Eric Isselee; 73*r*: Denis Babenko; 74: Eric Isselee; 75*l*: wavebreakmedia; 75*r*: Joop Snijder Photography; 76*l*: Monika Wisniewska; 76*r*: WilleeCole Photography; 77*a*: rebecca ashworth; 77*b*: wavebreakmedia; 78*l*: Eric Isselee; 78*c*: Peter Waters; 79*a*: Anke van Wyk; 79*bl*: P.Burghardt; 79*br*: Eric Isselee; 80: Eric Isselee; 81: eClick; 82: Nata Sdobnikova; 83*ar*: Ermolaev Alexander; 83*b*: Vitaly Titov & Maria Sidelnikova; 84: sbko; 85*l*: Lenkadan; 85*r*: WilleeCole Photography; 86: shutterstock; 87*a*: Monika Wisniewska; 87*b*: bibiphoto; 88: Eric Isselee; 89*ar*: Andrey_Popov; 89*b*: Eric Isselee; 90: Javier Brosch; 91: Keattikorn; 92: Ermolaev Alexander; 93: Tatiana Katsai; 94: MilsiArt; 95: Javier Brosch; 96: KariDesign; 97: Susan Schmitz; 98: siamionau pavel; 99*l*: Jaimie Duplass; 99*c*: Liliya Kulianionak; 99*r*: Susan Schmitz; 100: michaeljung; 101: WilleeCole Photography; 102: Javier Brosch; 103*l*: Marina Jay; 103*r*: Susan Schmitz; 104: Eric Isselee; 105: cynoclub; 106: Ivonne Wierink; 107: WilleeCole Photography; 107*ar*: Loskutnikov; 108*ar*: Loskutnikov; 108*bl*: Javier Brosch; 108*c*: Africa Studio; 109*l*: Erik Lam; 109*br*: Africa Studio; 110: Vitaly Titov & Maria Sidelnikova; 111*a*: zeljkodan; 111*b*: Vitaly Titov & Maria Sidelnikova; 112: mariait; 113*ar*: AnetaPics; 113*bl*: Eric Isselee; 113*br*: Artem Furman; 114: Jeroen van den Broek; 115*a*: Monkey Business Images; 115*b*: Anna Hoychuk; 116*l*: Len44ik; 116*r*: sergeevspb; 117: margouillat photo; 118: margouillat photo; 119*l*: Baevskiy Dmitry; 119*ar*: Tatiana Popova; 120: Jari Hindstroem; 121: Susan Schmitz; 122*l*: Eldad Carin; 122*ar*: cretolamna; 122*br*: Erik Lam; 123*b*: Erik Lam; 123*ar*: Ugorenkov Aleksandr; 124: Eric Isselee; 125*cl*: Mageon; 125*r*: Shevs; 126: Elliot Westacott; 127*bl*: Mona Makela; 127*r*: Anna Hoychuk; 128: Africa Studio; 129*ar*: Eric Isselee; 129*b*: iko; 130*a*: AnetaPics; 130*b*: AnetaPics; 131*al*: AnetaPics; 131*bl*: AnetaPics; 131*r*: Anatoly Tiplyashin; 132: wavebreakmedia; 133*ar*: Igor Normann; 133*b*: Eric Isselee; 134: tobkatrina; 135*a*: Marcella Miriello; 135*b*: fotostory; 136: Dorottya Mathe; 137*l*: Javier Brosch; 137*ar*: sandra zuerlein; 138*b*: Erik Lam; 138*a*: cretolamna; 139*al*: Ksenia Raykova; 139*br*: Ermolaev Alexander; 140: GWImages; 141*a*: vvvita; 141*b*: Dirima; 142: Wilson's Vision; 143*a*: think4photop; 143*b*: plavevski; 144*l*: Annette Shaff; 144*r*: Annette Shaff; 145*l*: Dorottya Mathe; 145*ar*: think4photop; 146: sav-in; 147: a katz; 148: cynoclub; 149: Miriam Doerr; 150-151: cellistka; 151*r*: Venus Angel; 152*l*: Eric Isselee; 152*c*: Aaron Amat; 153*ar*: Dennis W. Donohue; 153*b*: cynoclub; 154: Sparkling Moments Photography; 155*a*: Mikel Martinez; 155*b*: Dennis W. Donohue; 156*l*: Eric Isselee; 156*ar*: Mackland; 157: Eric Isselee; 158: Erik Lam; 159*ar*: Andraž Cerar; 159*b*: Dziurek; 160: Eric Isselee; 161*ar*: Mackland; 161*b*: SueC; 162*l*: Venus Angel; 162*r*: Claire McAdams; 163*ar*: Milan Vachal; 163*b*: Mackland; 164: Margo Harrison; 165: francesco de marco; 166-167: Marcel Jancovic; 168: Richard Chaff; 169*a*: Magnum Johansson; 169*cr*: Mageon; 169*b*: Barna Tanko; 170: Ermolaev Alexander; 171*al*: cynoclub; 171*r*: cynoclub; 172: Lisa F. Young; 173*ar*: Alexia Khruscheva; 173*b*: Eric Isselee; 174: Erik Lam; 175*a*: vgm; 175*b*: ataglier; 176: Jeffrey B. Banke; 177*a*: Tom Feist; 177*b*: gorillaimages; 178*l*: EpicStockMedia; 178*r*: Steve Collender; 179*ar*: rSnapshotPhotos; 179*b*: Deborah Kolb; 180: Marcel Jancovic; 181*ar*: Lee319; 181*b*: Anke van Wyk; 182; Anke van Wyk; 183*ar*: JASPERIMAGE; 183*b*: gillmar; 184: Baevskiy Dmitry; 185*l*: Baevskiy Dmitry; 185*ar*: glen gaffney; 186: eAlisa; 190: Mat Hayward.

Quercus

New York • London

Copyright © 2015 Quercus Editions Ltd
First published in the United States by
Quercus in 2015

Any member of educational institutions
wishing to photocopy part or all of the work
for classroom use or anthology should send
inquiries to permissions@quercus.com.

ISBN 978-1-62365-487-0

Library of Congress Control Number:
2014948255

Distributed in the United States and Canada by
Hachette Book Group
1290 Avenue of the Americas
New York, NY 10104

Manufactured in China

10 9 8 7 6 5 4 3 2 1

www.quercus.com

Text by David Alderton
Edited by Philip de Ste. Croix
Designed by Sue Pressley and Paul Turner,
Stonecastle Graphics Ltd